Autodesk Civil 3D 2024 from Start to Finish

Build scalable, real-world infrastructure projects using Civil 3D's full design and modeling toolkit

Stephen Walz

Tony Sabat

BIRMINGHAM—MUMBAI

Autodesk Civil 3D 2024 from Start to Finish

Group Product Manager: Rohit Rajkumar

Publishing Product Manager: Kaustubh Manglurkar

Content Development Editor: Abhishek Jadhav

Technical Editor: Simran Ali

Copy Editor: Safis Editing

Project Coordinator: Sonam Pandey

Proofreader: Safis Editing

Indexer: Rekha Nair

Production Designer: Alishon Mendonca

Marketing Coordinator: Nivedita Pandey

First published: April 2023

Production reference: 2010725

Published by Packt Publishing Ltd.

Livery Place

35 Livery Street

Birmingham

B3 2PB, UK.

ISBN 978-1-80323-906-4

www.packtpub.com

I want to express my heartfelt appreciation to all of my colleagues, mentors, friends, and family, for all your guidance and encouragement throughout my career, with a special shoutout to my wife, Danna Walz, and daughters, Alexis and Addison, for all of your continuous love and support, and inspiration for me to be the best version of myself every day!

- Stephen Walz

Thank you to all of my colleagues and friends in the industry that helped me pursue and grow my love for technology. Also to my wife, Joy, thank you for your support and patience throughout this process and throughout my career to keep me driving forward and enjoying every step along the way.

- Tony Sabat

Contributors

About the authors

Stephen Walz has worked with a multitude of design, collaboration, and visualization platforms supporting the AEC industry since early 2003. His primary focus has been on the civil and environmental engineering side, where he has held varying levels of design support and CAD/BIM/CIM management roles.

Currently, Stephen is HDR's digital design lead for civil infrastructure, where he works with their BG leadership, technical leadership, vendors, and ITG to evaluate and implement new technologies, workflows, and adoption strategies supporting the AECOO industry, build awareness and drive consistency with how HDR's teams leverage various tools and platforms, and identify ways to build skill sets across HDR around each of the platforms and technologies being leveraged.

Furthermore, Stephen is an active member of the following:

- AUGIWORLD Magazine as the BIM/CIM content manager
- buildingSMART International as a member of the Education and Professional Certification Committee within buildingSMART's US chapter
- The **National Institute of Building Sciences** (**NIBS**) as an NCS V7 Project Committee member

Tony Sabat is a consultant and advisor focusing on improving the built environment. He primarily develops innovative and disruptive technologies and processes with many teams, varying from early-stage startups to Fortune 500 companies. Tony partners with companies to develop their digital transformation strategies, as well as implement and sustain emerging technologies, such as reality data modeling, virtual construction simulation, and distributed ledger technology. He has also spoken around the world and written for many organizations about emerging technology in the built environment.

Tony began his career in the early iterations of building information modeling and worked on integrating these concepts and technologies into the infrastructure space. Tony focuses primarily on reality capture, building information modeling, digital twin strategies, as well as virtual design and construction.

About the reviewer

Justin Brooks, PE, PMP has been an avid user of Civil 3D, along with other Autodesk products, during his career of over 20 years in civil engineering. He has an undergraduate and graduate degree in civil engineering and an associate's degree in CAD as well.

He has held multiple roles within the civil engineering industry, from designer all the way up to project engineer and project manager, with his current role being that of design technology manager with Civil & Environmental Consultants, Inc. Along with his time in the industry, he has also spent the latter part of his career in post-secondary education as both a professor and curriculum developer in construction management and technology programs.

Table of Contents

3

Part 2: Designing and Modeling with Civil 3D from Scratch

4

5

6

Surfaces - The First Foundational Component to Designs within Civil 3D   119

7

Alignments - The Second Foundational Component to Designs within Civil 3D   145

8

Profiles - The Third Foundational Component to Designs within Civil 3D   175

Part 3: Leveraging Design-Specific Tool Belts

9

10

11

12

Part 4: Advanced Capabilities with Civil 3D

13

14

Preface

Autodesk Civil 3D connects to an entire network of other software to take your project farther than you would imagine for civil engineering design. You will be able to put your knowledge to work with this practical guide on civil engineering design. This book provides a hands-on approach to implementation and associated methodologies that will have you and your team functional and productive in no time.

This book covers all the major features Civil 3D has to offer, whether surface development, intelligent utility design, or dynamic display work for creating smart documentation. You'll learn how best to configure and manage your civil engineering designs by successfully leveraging all that Autodesk Civil 3D has to offer. You'll then learn how to practically apply the tools and modeling techniques available within the software.

By the end of this book, you will have a thorough understanding of Autodesk Civil 3D, along with its partner programs, and be able to strategize how best to utilize it on your next project.

Who this book is for

This book is for civil engineers, environmental engineers, surveyors, civil designers, civil technicians, Civil 3D professionals, and InfraWorks professionals looking to understand how to best leverage Civil 3D in their everyday designs.

You'll need to have a very basic understanding of civil engineering and surveying workflows, as well as a foundational understanding of Autodesk's AutoCAD, to make the most of this book. A basic understanding of surveying, civil/environmental engineering practices, and AutoCAD drafting knowledge is assumed.

What this book covers

Chapter 1, Introduction to Civil 3D, begins by exploring the program itself. This is where the main tools are, and you will familiarize yourself with the working space. We will explore the user interface and review where objects are stored as they are created.

Chapter 2, Setting up the Design Environment, reviews how work is logged and integrated as more complexity is added. We will explore how modeled components and geometry can be affected by dynamic display styles and how to create a new file for a design project.

Chapter 3, Sharing Data within Civil 3D, demonstrates the power of Civil 3D as a program, not only showing how to design a basic project in Civil 3D but also how to scale your workflows to include more team members for larger project sizes while maintaining efficiency.

Chapter 4, Configuring Survey Data with Civil 3D, contains all you need to know about bringing in existing survey data with proper formatting and how to establish an accurate depiction of a yet-to-be-designed project site, given that civil design always begins with existing conditions.

Chapter 5, Leveraging Points, Lines, and Curves, dives into the application and intersection of Civil 3D's tools, now that we understand the basic tools and methods, to better develop a confident infrastructure design. Civil 3D is a dynamic program that not only makes use of intelligent components and design processes but also integrated workflows that adjust when dependent parts of the design adjust or change, which results in less time spent recreating work or formatting when work is updated automatically.

Chapter 6, Surfaces - The First Foundational Component to Designs within Civil 3D, elaborates on one of the three foundational modeling elements, which is surface modeling, required for almost every civil engineering-related design.

Chapter 7, Alignments - The Second Foundational Component to Designs within Civil 3D, is all about the second foundational modeling element required, which is alignments, pertinent to almost every civil engineering-related design.

Chapter 8, Profiles - The Third Foundational Component to Designs within Civil 3D, explains the third and final foundational modeling element, which is profiles, as it applies to almost every civil engineering-related design.

Chapter 9, Land Development Tool Belt for Everyday Use, involves putting on our Land Development Tool Belt and jumping into the world of subdivisions, parks, and overall civil site design.

Chapter 10, Roadway Modeling Tool Belt for Everyday Use, has us put on our Roadway Modeling Tool Belt and jump into the world of transportation.

Chapter 11, Advanced Roadway Modeling Tool Belt for Everyday Use, continues to expand upon our Roadway Modeling Tool Belt and jumps into advanced applications to transportation.

Chapter 12, Utility Modeling Tool Belt for Everyday Use, walks through a final specialized set of tools, the Utility Tool Belt, and jumps into the world of utility infrastructure, which ultimately services our communities and land.

Chapter 13, Section Creation and Analysis, reviews how to create, display, and manage Section Views within Civil 3D.

Chapter 14, Automating Sheet Creation, teaches you how to automate sheet creation and migrate your design from a modeled state to a true design deliverable.

To get the most out of this book

Software/hardware covered in the book	Operating system requirements
Autodesk AutoCAD	Windows or macOS
Autodesk Civil 3D	Windows or macOS

If you are using the digital version of this book, we advise you to use the datasets available as you work through each chapter. Instructions are available within the chapters that detail whether readers will need to create a new drawing from scratch or open a file that has already been prepared. Doing so will help you follow the steps outlined.

Disclaimer

To ensure that a timely publication of this book's content is in close alignment with Autodesk's product release cycle, you may notice as you progress through our Civil 3D 2024 learning journey that there are several images displayed in referenced Figures that depict an earlier release of Civil 3D. It's important to note that features and workflows discussed throughout this book are still relevant and applicable in Autodesk's Civil 3D 2024 product and in which we will be applying throughout our learning journey.

Download the exercise files

You can download the exercise files for this book at `https://packt.link/UoiPn`

Download the color images

We also provide a PDF file that has color images of the screenshots and diagrams used in this book. You can download it here: `https://packt.link/2lHjT`.

Conventions used

There are a number of text conventions used throughout this book.

`Code in text`: Indicates code words in text, database table names, folder names, filenames, file extensions, pathnames, dummy URLs, user input, and Twitter handles. Here is an example: "Please note that if your design references metric units, you can select the `_Autodesk Civil 3D (Metric) NCS.dwg` drawing template to ensure that correct units are applied to your drawing."

Bold: Indicates a new term, an important word, or words that you see onscreen. For instance, words in menus or dialog boxes appear in **bold**. Here is an example: "Any time you create a new drawing by going to **Menu Browser** and selecting **New** | **Drawing**, a dialog box will appear for you to select the drawing template."

> **Tips or important notes**
> Appear like this.

Get in touch

Feedback from our readers is always welcome.

General feedback: If you have questions about any aspect of this book, email us at `customercare@packtpub.com` and mention the book title in the subject of your message.

Errata: Although we have taken every care to ensure the accuracy of our content, mistakes do happen. If you have found a mistake in this book, we would be grateful if you would report this to us. Please visit `www.packtpub.com/support/errata` and fill in the form.

Piracy: If you come across any illegal copies of our works in any form on the internet, we would be grateful if you would provide us with the location address or website name. Please contact us at `copyright@packt.com` with a link to the material.

If you are interested in becoming an author: If there is a topic that you have expertise in and you are interested in either writing or contributing to a book, please visit `authors.packtpub.com`

Share Your Thoughts

Once you've read *Autodesk Civil 3D from Start to Finish*, we'd love to hear your thoughts! Scan the QR code below to go straight to the Amazon review page for this book and share your feedback.

`https://www.amazon.in/review/create-review/error?asin=1803239069`

Your review is important to us and the tech community and will help us make sure we're delivering excellent quality content.

Download a free PDF copy of this book

Thanks for purchasing this book!

Do you like to read on the go but are unable to carry your print books everywhere?

Is your eBook purchase not compatible with the device of your choice?

Don't worry, now with every Packt book you get a DRM-free PDF version of that book at no cost.

Read anywhere, any place, on any device. Search, copy, and paste code from your favorite technical books directly into your application.

The perks don't stop there, you can get exclusive access to discounts, newsletters, and great free content in your inbox daily

Follow these simple steps to get the benefits:

1. Scan the QR code or visit the link below

https://packt.link/free-ebook/9781803239064

2. Submit your proof of purchase
3. That's it! We'll send your free PDF and other benefits to your email directly

Part 1: Getting Acquainted with Civil 3D and Starting Your Next Project for Success

Civil 3D is intelligent civil design software with dynamic elements that can link together throughout the course of your design and its refinement.

The following chapters are included in this section:

1

Introduction to Civil 3D

Autodesk's Civil 3D has been the leading design authoring tool of choice supporting civil engineering design applications since the early 2000s. Designers working with Civil 3D are able to put their knowledge to work with this practical guide to civil engineering design. This textbook is intended for civil designers with a basic understanding of Civil 3D and its workflows.

With this book, you will learn how best to configure and manage your civil engineering designs by successfully leveraging all that Autodesk Civil 3D has to offer.

We will begin by exploring best practices and real-world applications that will elevate your designs with Civil 3D. Autodesk Civil 3D connects to an entire network of other platforms and technologies to take your project further than you would imagine for civil engineering designs. By the end of this book, you will have a thorough understanding of Autodesk's Civil 3D and will be able to strategize how best to utilize it in your next project.

In this chapter, we will provide some background and history to help you realize the benefits of leveraging Autodesk's Civil 3D and understand its place within the **Architeture, Engineering and Construction (AEC)** industry. We will then familiarize ourselves with the overall user interface to understand all the tools and functionality available within Autodesk's Civil 3D, allowing you to maintain a high level of efficiency as you work through your design(s). Finally, to gain all of the benefits that Civil 3D has to offer and maintain a high level of efficiency and consistency, you're going to learn how to leverage as much of Civil 3D's dynamic capabilities as possible to support your **Building Information Modeling (BIM)** design(s).

So, in this chapter, we will cover the following topics:

- What is Civil 3D?
- The user interface
- Understanding Civil 3D elements

Technical requirements

It's important to note that Autodesk's Civil 3D can often be very taxing to your computer. There is a lot of processing that goes on with modeled design elements, even in the background, that enables the dynamic (connected) capabilities to occur throughout the BIM design life cycle. In turn, there are many technical requirements that need to be considered to allow Autodesk's Civil 3D to operate at its full potential. Here, we'll review the minimum requirements that Autodesk recommends, with a few of my suggestions added to increase efficiency and speed throughout the BIM design process:

- Operating system – 64-Bit Microsoft Windows 10
- Processor – 3+ GHz
- Memory – 16 GB RAM (I suggest going with either 64 GB or 128 GB)
- Graphics card – 4 GB (I suggest going with 8+ GB)
- Display resolution – 1980 x 1080 with True Color
- Disk space – 16 GB
- Pointing device – MS-Mouse compliant

What is Civil 3D?

We've all heard the term BIM being thrown around for quite some time now. In general, BIM is the process typically applied to and associated with intelligently and dynamically designing an actual building or structure, where architects/engineers and designers can visualize and anticipate the true constructability of a project.

On the survey and civil engineering side, we've been generating, designing, and modeling everything outside buildings/structures in a 3D environment for just as long, if not longer, than the term *BIM* has been around. However, for one reason or another, these designs have not typically been viewed as BIM by the vast majority.

This is where Autodesk's Civil 3D comes into the picture. Civil 3D is essentially a BIM design authoring tool that fully supports both surveying and civil engineering designs. When utilized the right way, we have the power to create and manage intelligently designed models, perform cost estimations, interference checks/clash detections, prepare our models for construction sequencing, and even perform asset management integrations well beyond design.

It's important to note that Autodesk's Civil 3D is considered a vertical application of Autodesk's AutoCAD. A vertical application in this context essentially means that Civil 3D is built on top of AutoCAD, thereby still providing a lot of 2D-focused tools and functionality. To make the transition from 2D to 3D, we'll need to realize the separation of where the content resides and how the application of tools differs in each world. Throughout the course of this book, I will cover both 2D and 3D applications of design intent to provide a bit more insight as to how and when to apply each of them.

Historically, it is always recommended that anyone looking to fully utilize Civil 3D should have a basic understanding of AutoCAD to establish that foundation. Depending on the individual learning, that concept is more frequently dismissed as it has the potential to root bad design habits in those not willing to fully embrace the 3D world. As Civil 3D continues to evolve and become more of a model-centric focused design authoring tool, the industry has a much better opportunity to fully embrace the concept of BIM and design with intelligence from the beginning.

What types of BIM projects can Civil 3D manage?

When trying to determine how Civil 3D can best be utilized from a design standpoint, we're going to break down the tools and functionality into the following three main design categories or markets: land development, utilities, and transportation. In addition to the three main design categories, it's important to realize all projects, whether building or infrastructure, rely upon existing conditions, and Civil 3D is a great partner for hosting these parameters as well as modeling new ones for a complementary design to the interior of the structure. This added benefit is heavily relied on and supported in the surveying industry as well.

In this course, we'll cover all four of these categories in great length as follows:

- From a surveying perspective, we'll cover tools that allow for full integration of survey databases, field books, existing conditions modeling, and even point cloud integration from drone and laser scan technology. Any BIM design being generated can only be as accurate as the survey and existing conditions model, making these early steps critical to successful design downstream.

- From a land development perspective, we'll cover methods that allow a wide array of civil site designs including subdivisions, residential/commercial sites, parks, and even environmental designs.

- From a utility perspective, we'll cover tools and functionality that support storm drainage, sanitary sewer, waterline, and natural gas designs.

- From a transportation perspective, we'll cover tools and functionality that support roadway designs, rehabilitations, and access roads. Additionally, we'll cover tips and tricks on how some of these tools can be applied in civil site designs and even stream and restoration projects.

Understanding real-world applications of any new technology solution is critical from an overall adoption and implementation standpoint. Throughout this book, especially as we dive deeper into the specific tools and functionality of Autodesk's Civil 3D, we'll begin to understand how and why it is viewed as a leading design authoring tool supporting the AEC industry.

Let's go ahead and jump into the program to begin to familiarize ourselves with the user interface to get an idea of where to locate all of the tools and functionality available to us.

Exploring the Civil 3D user interface

With all the introductions and background out of the way, let's go ahead and jump into Autodesk Civil 3D and get acquainted with this BIM platform. When Autodesk Civil 3D is opened up or launched, you will be presented with the **Start** tab/screen, similar to that shown in *Figure 1.1*:

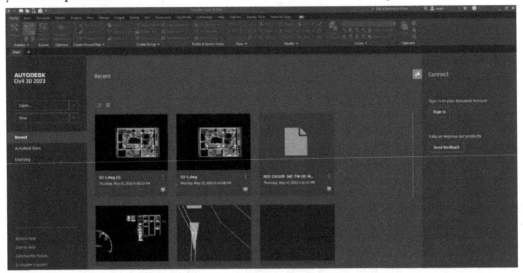

Figure 1.1 – User interface when Autodesk's Civil 3D is launched

If we then select the + sign next to the **Start** tab, we will see we have essentially created our first drawing and will be presented with a screen similar to that shown in *Figure 1.2*:

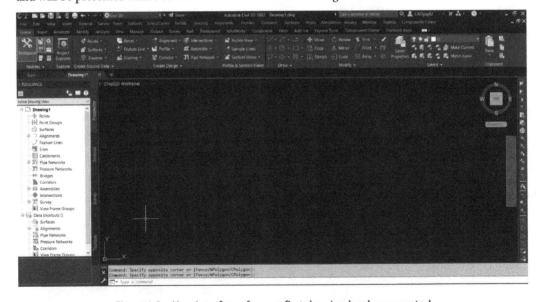

Figure 1.2 – User interface after our first drawing has been created

Along the very top of the program, you are presented with three levels of tools and functionality, as shown in *Figure 1.3*:

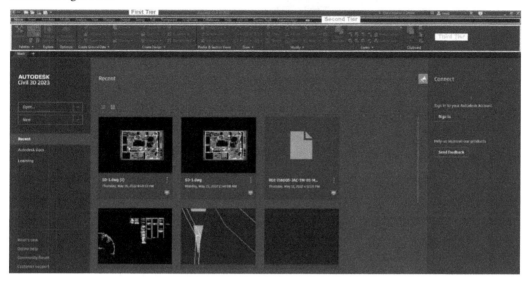

Figure 1.3 – Top three levels of tools and functionality within the user interface

Each of these levels provides varying access to the tools and functionality available within Autodesk's Civil 3D platform. Much of this should be a bit clearer by the end of the section, as we dive into more detail about what is available at each level.

The first tier of the ribbon

In the first level at the very top, reading from left to right, we have access to the **Menu** browser, **Quick Access Toolbar, Workspace and Quick Access Toolbar Customizations, Application and Drawing currently visible in the Drawing area, Quick Search Access to Autodesk's Civil 3D Help, Autodesk Account Access, Overall Autodesk's Civil 3D Help**, and **Quick program visibility access**.

Starting with the **Menu** browser, which is indicated by the Civil 3D icon in the top-left corner of the program, you will be presented with a multitude of options that will provide some very high-level drawing management and functionality. Select the icon by left-clicking on your mouse (refer to *Figure 1.4* for a detailed view of the **Menu** browser):

Figure 1.4 – Detailed view of the Menu browser

Running through the options presented along the left side of the **Menu** browser from the top to the bottom, the following list details what each area provides:

- **New**: This selection allows end users to create new drawings and sheet sets (more on sheet sets in *Part 4* of the book).

- **Open**: This selection allows end users to open existing drawings (both locally or web and mobile-based), access **Sheet Set Manager**, import DGN files (Microstation format), import **Industry Foundation Class** (**IFC**) files, and open sample files (both locally or from online).

- **Save**: This selection allows end users to perform a quick save of their current drawing.

- **Save As**: This selection allows end users to save their current drawing as another drawing (both locally or web and mobile-based), drawing template file, drawing standards file, other formatted files (that is, DWG, DWDT, WS, and DXF), save the layout as a drawing file, and DWG Convert which will allow us to save to a DWG format that is compatible with earlier releases of Civil 3D).

- **Export**: This selection allows end users to export their current drawing to a DWF, DWFx, 3D DWF, another DWG version, DGN, DXF, PDF, IFC, and other file formats (that is, WMF, SAT, STL, EPS, DXX, BMP, IGES, and IGS).

- **Publish**: This selection allows end users to access the **Send to Print** service, **Archive**, **eTransmit**, **Email**, and **Share View**.

- **Print**: This selection allows end users to access **Plot**, **Batch Plot**, **Plot Preview**, **View Plot and Publish Details**, **Page Setup**, **3D Print**, **Manage Plotter**, **Manage Plot Styles**, and **Edit Plot Style Tables**.

- **Suite Workflows**: This selection allows end users to access **Workflow Manager**, a dedicated area for common automated tasks.

- **Drawing Utilities**: This selection allows end users to access **Drawing Properties, DWG Compare, Drawing Settings, Units, Audit, Status, Purge, Recover, Open the Drawing Recovery Manager**, and **Update Block Icons**.

- **Close**: This selection allows end users to close the current drawing or all drawings.

Moving onto **Quick Access Toolbar**, we have a lot of the same options available to us as are within the **Menu** browser. However, instead of a drop-down menu like that of the **Menu** browser, we quick-select icons of tools and functionality that are typically common to everyday workflows:

Figure 1.5 – Quick Access Toolbar

Running through the stock, out-of-the-box **Quick Access** icons from left to right (as shown in *Figure 1.5*), we have the following:

- **New**
- **Open**
- **Save**
- **Open from Web & Mobile**
- **Save to Web & Mobile**
- **Plot** (print)
- **Undo**
- **Redo**

> **Note**
>
> Next to the **Undo** and **Redo** icons in **Quick Access Toolbar**, there are *down* arrows that will allow you to select the number of steps for which you would like to perform that operation on your drawing. By simply selecting the icon instead of the *down* arrow, you will perform that operation once for each click of the mouse. If you are intending to perform that operation multiple times, I recommend using the *down* arrow next to the icon to get the drawing back to the state that you are expecting.

If there are additional tools that are typical to your everyday workflow, there is a way to add them to your toolbar by simply right-clicking on the tool in the ribbon and selecting **Add to Quick Access Toolbar**.

Next, we have the **Workspace and Customized Quick Access** toolbar. Here you can change the tools available in the ribbons on the third and fourth levels depending on the type of design and analysis you are looking to perform within the current drawing. To change from one workspace to another, you can select the *down* arrow icon next to the workspace name and select a different one as appropriate (see *Figure 1.6*).

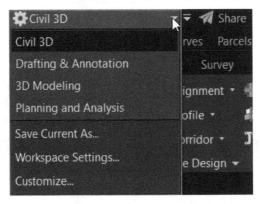

Figure 1.6 – Workspace selections

Workspaces displayed and available include the following:

- **Civil 3D**: Tools displayed in the ribbon will provide you with the bulk of the tools necessary to perform civil engineering design and analysis.

- **Drafting & Annotation**: Tools displayed in the ribbon will provide you with the basic 2D drafting tools and functionality.

- **3D Modeling**: Tools displayed in the ribbon will provide you with the basic 3D modeling tools and functionality. It's important to note that 3D modeling capabilities within this display are not Civil 3D-related but are more intended for basic AutoCAD functionality.

- **Planning and Analysis**: Tools displayed in the ribbon will provide you with additional mapping and analysis tools. These tools come in handy when connecting to ArcGIS data.

- **Save Current As…**: Provides the ability to save any customizations you have made to a customized workspace that can be made current later on.

- **Workspace Settings…**: Selecting this option will pull up another dialog box that will allow you to define which workspaces are available, as well as the display order of the workspaces listed.

- **Customize…**: Selecting this option will pull up another dialog box that will allow you to customize any of your workspaces if necessary. This will come in handy if you intend to create custom tools, layouts, and so on specific to your workflow.

Directly next to the workspace selections is another *down* arrow with a horizontal line above it. If you select this, you will be presented with a drop-down menu that will provide the ability to customize **Quick Access Toolbar** and **Tools Available** within the overall ribbon interface (refer to *Figure 1.7*):

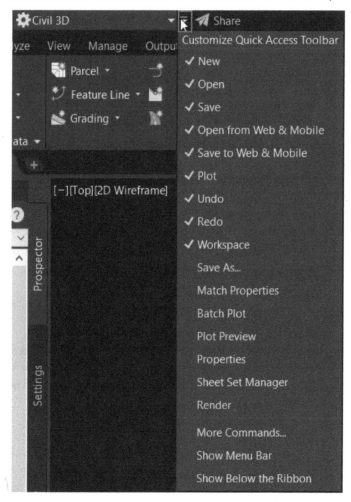

Figure 1.7 – Customize Quick Access Toolbar

Those items that are checked will be displayed in **Quick Access Toolbar** in your current view. You'll notice that all of those checked in *Figure 1.7* were those that were displayed when we were reviewing **Quick Access Toolbar** (*Figure 1.5*). The remaining tools listed in the top section are fairly straightforward and are typical commands and functionality that are relied on for a lot of workflows, hence the reason why they are listed as optional defaults in the list.

The last option in **Quick Access Toolbar** is an icon that looks like a paper airplane (see *Figure 1.8*). Selecting this icon will allow you to share the current drawing via a link to a web-based version. By default, the drawing is only available for 7 days.

Figure 1.8 – Share the drawing with a link

Next, we have **Application and Drawing** currently visible in the drawing area (refer to *Figure 1.9*). As you open or create and save new drawings, the drawing name displayed in this location will update accordingly.

Figure 1.9 – Application and Drawing

Then, we move on to **Quick Search Access to Autodesk's Civil 3D Help** (refer to *Figure 1.10*). Any time you find yourself needing some quick help on a specific command, simply type the name of the command in this location and click on the magnifying glass icon next to it. After selecting the magnifying glass icon, another window will appear that contains the Civil 3D Help documentation and will present you with results that match the keyword or phrase you had typed into the field.

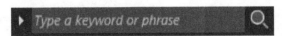

Figure 1.10 – Quick Search Access

Next up is **Autodesk Account Access**. Viewing from left to right, we first have **Account Login and Details** presented to you. When selecting the *down* arrow icon in the view (refer to *Figure 1.11*), you will have the ability to log out, view account details, explore purchase options, and manage licenses.

Figure 1.11 – Autodesk Account Access

Next to **Autodesk Account Access** (refer to *Figure 1.12*), we are presented with the option to connect to the Autodesk App Store (indicated by the shopping cart icon) and to connect to Civil 3D user networks via Facebook and Twitter:

Figure 1.12 – Autodesk App Store and user group networks

Continuing on, we have the overall **Autodesk Civil 3D Help** access, which is identified by the question mark icon (see *Figure 1.13*):

Figure 1.13 – Autodesk Civil 3D Help

And finally, we come to the **quick program visibility access**, where we have the ability to minimize the window of the program, restore and maximize the window of the program, and close the entire program:

Figure 1.14 – Quick program visibility

With the contents of the first tier behind us, let's jump into the second tier.

The second tier of the ribbon

In the second level of the ribbon interface, we have the tabs that essentially group all of the tools and functionality available within Autodesk's Civil 3D (refer to *Figure 1.15*). This grouping is meant to be an easy way to quickly access tools as you need them, instead of having to search through the drop-down menus for tools as needed.

Figure 1.15 – Second tier

Starting with the **Home** tab (refer to *Figure 1.16*), we have quick access to many of the major design tools and functionality that we will be leveraging through the BIM design life cycle:

Figure 1.16 – Home

Next up is the **Insert** tab (refer to *Figure 1.17*), where we can quickly make connections to other types of files for reference purposes. Based on the type of file(s) being referenced, we can either keep these as static overlays being displayed or can make them live elements within our drawing area:

Figure 1.17 – Insert

Next up is the **Annotate** tab (refer to *Figure 1.18*), where we can quickly add labels and text to various design elements:

Figure 1.18 – Annotate

Next up is the **Modify** tab (refer to *Figure 1.19*), where we can quickly adjust design elements within the drawing:

Figure 1.19 – Modify

Next up is the **Analyze** tab (refer to *Figure 1.20*), where we can quickly access various reporting and computational tools and functionality available to us:

Figure 1.20 – Analyze

Next up is the **View** tab (refer to *Figure 1.21*), where we can quickly adjust our display settings of modeled elements in our drawing area:

Figure 1.21 – View

Next up is the **Manage** tab (refer to *Figure 1.22*), which, in a way, is more of a parking lot that provides quick access to a multitude of options, such as connecting to individual design elements from other Civil 3D files, performing customizations to the overall user interface, and file cleanup options:

Figure 1.22 – Manage

Next up is the **Output** tab (refer to *Figure 1.23*), where we can quickly export our drawing and individual design elements to a variety of formats:

Figure 1.23 – Output

Next up is the **Survey** tab (refer to *Figure 1.24*), where we can quickly access various survey and mapping tools to support our BIM designs:

Figure 1.24 – Survey

Next up is the **Rail** tab (refer to *Figure 1.25*), where we can quickly access BIM design capabilities related to Rail:

Figure 1.25 – Rail

Next up is the **Transparent** tab (refer to *Figure 1.26*), where we can quickly access functionality that can be applied while performing various commands throughout our BIM design:

Figure 1.26 – Transparent

Next up is the **InfraWorks** tab (refer to *Figure 1.27*), where we can quickly connect to InfraWorks by importing/exporting design elements generated within either the Civil 3D or InfraWorks platform:

Figure 1.27 – InfraWorks

Next up is the **Collaborate** tab (refer to *Figure 1.28*), where we can quickly share our drawings and BIM design elements in various cloud-based environments:

Figure 1.28 – Collaborate

Next up is the **Help** tab (refer to *Figure 1.29*), where we can quickly access additional support as needed:

Figure 1.29 – Help

Next up is the **Add-ins** tab (refer to *Figure 1.30*), where we can quickly access any add-ins that are integrated with Civil 3D:

Figure 1.30 – Add-ins

Next up is the **Express Tools** tab (refer to *Figure 1.31*), where we can quickly access various ad hoc tools:

Figure 1.31 – Express Tools

Next up is the **Featured Apps** tab (refer to *Figure 1.32*), where we can quickly access the Autodesk App Store to purchase and install leading add-ins being leveraged across the AEC industry:

Figure 1.32 – Featured Apps

And finally, located at the very end of this tier just to the right of the **Featured Apps** tab is an icon that will give you the ability to cycle through the visibility states of the ribbon (as shown in *Figure 1.33*):

Figure 1.33 – Ribbon visibility states

Going back to what I said about the **Menu** bar visibility, depending on preference, you have the ability to either toggle on or off the tabs and panels display as well. Seeing as how the tabs and panels do consume a bit more real estate, or area, at the top of the session, you can minimize, even temporarily, to increase the space available in the drawing area as well (more on that in a bit). Ultimately, it comes down to preference of access and is in your hands to decide.

With the contents of the second level behind us, let's jump into the third.

The third tier of the ribbon

In the third and final level of the ribbon interface, we have the panels that are categorized one more level down and provide access to the actual tools and functionality available within Autodesk's Civil 3D (refer to *Figure 1.34*):

Figure 1.34 – Ribbon panel

It's far more important, in my opinion, to fully understand the purpose and functionality of these tools when we review real-world design workflows and applications of tools. With that said, all of the tools and functionality within each of the tabs and panels will be covered in future sections and chapters within this book, so I won't bore you with too much detail at this time. Please refer to *Figures 1.15* to *1.33* for a high-level overview of what is included from a panel standpoint within each of the tabs.

Before moving on to the next area of the user interface, it is important to note that the ribbon we have been exploring is contextual, meaning it will change the commands that are visible on the ribbon depending on the command you are using or on what object is selected in the drawing. When an object is selected, the contextual ribbon will display commands directly related to the object you have selected in an attempt to expedite related workflows. Next, you can see the contextual ribbon in its default state, as shown in *Figure 1.35*:

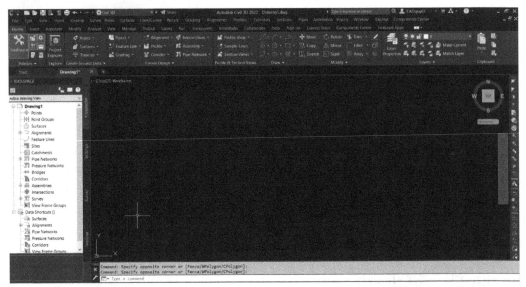

Figure 1.35 – Contextual ribbon when a command or tool is not active

The following figure shows when a tool or command *is active* or when modeled components are selected in the drawing area:

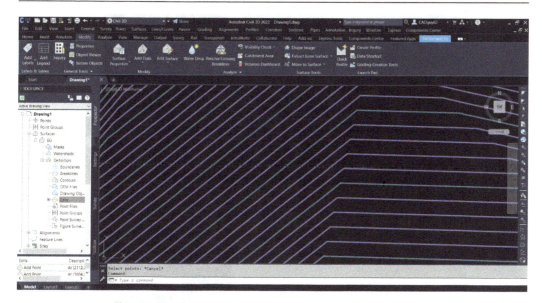

Figure 1.36 – Contextual ribbon when a command or tool is active

The major differences between the displays are the tools and commands that will be made available above your drawing area. The intent of this contextual ribbon is to narrow down the tool selections to what can actually be performed with the identified objects. This eliminates all kinds of unnecessary time trying to remember where to locate tools and ultimately makes your work much easier.

With the contents of the third level behind us, let's move on to the Toolspace.

Toolspace

The next part of the user interface that we'll review is **Toolspace**. This can be accessed by going to the **Home** tab and selecting the **Toolspace** icon in the **Palettes** panel. Once activated, you'll notice **TOOLSPACE** being displayed along the left-hand side of your session (refer to *Figure 1.37*):

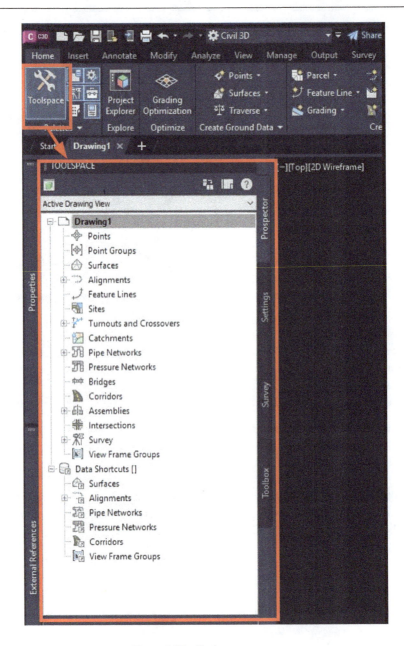

Figure 1.37 – Toolspace access

Within **TOOLSPACE**, you are presented with a ton of file and design management tools and functionality that will really drive your BIM design and the way modeled geometry and components are displayed. Along the right-hand side of your **TOOLSPACE** area, you'll notice four individual tabs that will give you access to each type of functionality, **Prospector**, **Settings**, **Survey**, and **Toolbox** (refer to *Figure 1.38*):

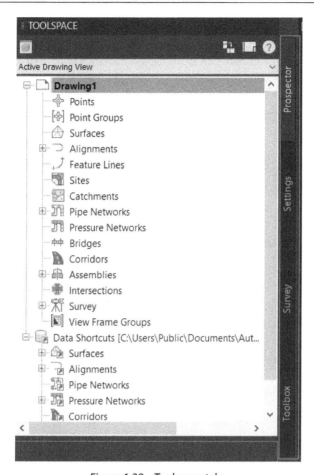

Figure 1.38 – Toolspace tabs

Let's look at each type of functionality:

- The **Prospector** tab provides you with the ability to create, modify, analyze, and manage modeled geometry and components within your drawing(s) open in the current session, as well as link modeled geometry and components from other files via **Data Shortcuts Manager**.

- The **Settings** tab provides you with the ability to create, modify, and manage settings specific to the drawing(s) open in the current session.

- The **Survey** tab provides you with the ability to connect to survey databases and manage configurations to streamline the integration of data within Autodesk's Civil 3D platform.

- The **Toolbox** tab provides you with additional reporting and analysis tools that work well for gathering information within the current drawing and exporting data to various reporting formats.

With the contents of the toolspace reviewed, let's now jump into the drawing area.

Drawing area

Moving on, the next area is the drawing area, which is fairly straightforward and self-explanatory. This area is reserved for your actual BIM design modeling, annotating, and so on. All your work will be displayed in this area and will appear as shown in this screenshot:

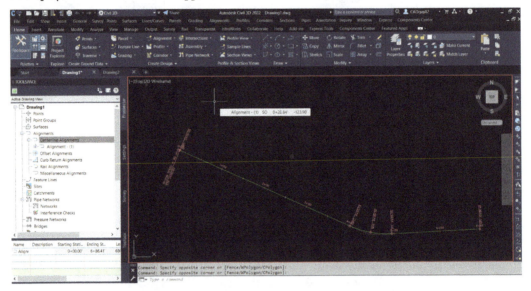

Figure 1.39 – Drawing area

You'll notice, however, a few different icons within your drawing area that are worth pointing out and providing a little more detail on. In the upper left-hand corner, you'll notice a few view descriptions, similar to [-] [Top] [2D Wireframe]. When you select each of these fields, you are presented with the ability to further manage and/or change the display of all elements displayed in the drawing area. If you select the first field, [-], you are presented with a few options to customize the drawing area, or model space viewport (shown in *Figure 1.40*):

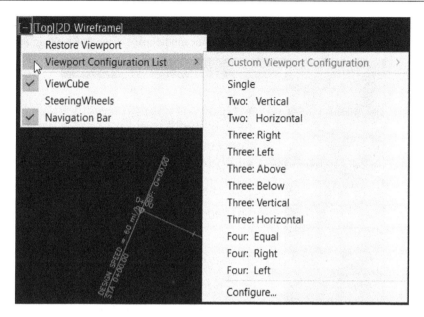

Figure 1.40 – Customize model space viewport

If you select the second field, **[Top]**, you are presented with a few options to customize the orientation of the view being displayed within your model space viewport (shown in *Figure 1.41*):

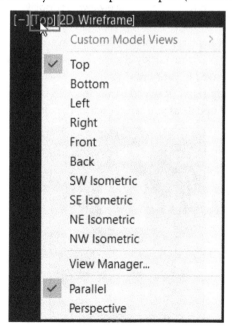

Figure 1.41 – Model space viewport view orientation

If you select the third field, [**2D Wireframe**], you are presented with the ability to change the overall display of your model geometry and components within the model space viewport (shown in *Figure 1.42*):

Figure 1.42 – Model space viewport display

Let's move over to the upper right-hand corner of your drawing area. This is where you'll find **ViewCube** (refer to *Figure 1.43*), which allows you to quickly change the **user coordinate system** (**UCS**) and view orientation of components displayed in your drawing area. By selecting any of the lettered directions or the rotation options in the top right, or selecting any location on the square/cube located in the center, the view orientation will update as you select each of the options.

Figure 1.43 – ViewCube

Additionally, if you select the *down* arrow toward the bottom right of **ViewCube**, you are presented with some additional view settings, along with display options for **ViewCube** itself (see *Figure 1.44*):

Figure 1.44 – ViewCube settings

Lastly, just underneath **ViewCube**, you'll notice a **Quick View Access Bar**, as shown in *Figure 1.45*. On this bar, we have the ability to quickly pull up the **Full Navigation Wheel** that will display right next to your mouse, pan across your drawing (represented by the hand icon), zoom in/out of your view, orbit/rotate your view along an axis, and **ShowMotion**, which allows you to create animations.

Figure 1.45 – Quick View Access Bar

With the contents of the drawing area reviewed, let's move on to understanding the command line.

Command line

Last, but not least, we have the command line (refer to *Figure 1.46*). Any time you are in a command, and not presented with a dialog box, you will see the next steps for your command being displayed at the command line.

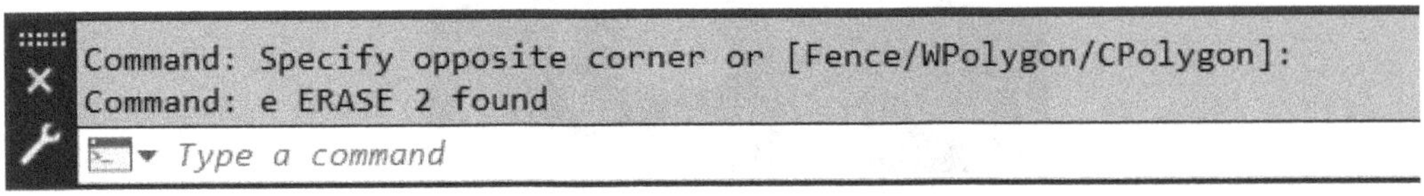

```
Command: Specify opposite corner or [Fence/WPolygon/CPolygon]:
Command: e ERASE 2 found
      ▼ Type a command
```

Figure 1.46 – Command line

It's important to keep an eye on the command line if you are in doubt as to what the next step of your command is. Further, if you're a "typer" like I am, this is also a great place to type any shortcuts of commands instead of trying to locate them within any of the menu dropdowns or sorting through the tools displayed in the ribbons. As an example, you can type the letters CO in the command line and hit *Enter* or the spacebar to initiate the Copy command, type the letter M and hit *Enter* or the spacebar to initiate the Move command, and so on.

Now that we have a high-level understanding of what tools are available to us and where they are located within the Autodesk Civil 3D environment, let's jump into understanding the dynamic nature of the platform to get a little more of an idea as to how best we can leverage it in a BIM design setting.

Understanding Civil 3D elements

Now that we're getting a little more familiar with the various access points to locate our tools, let's try to get a better understanding of the overall functionality and capabilities of Autodesk's Civil 3D. The way this BIM design authoring tool is built allows for multiple design geometry and components to be linked dynamically and, in a way, speak to each other. When one component is updated, it can potentially affect many other components that are built on top of it.

When you think about building a house, you start with a foundation, then you typically frame the house on top of the foundation, and then finish with the walls, roof, doors, windows, insulation, and so on. Thinking about how this workflow applies on the civil engineering side, we typically start with a survey, next we'll lay out our proposed site geometry as the framework, and then build everything else based on this framework (roadways, utilities, etc.).

Within Autodesk's Civil 3D, we can create the following intelligent components that have the potential to be linked dynamically to each other:

- Points
- Surfaces

- Sites

- Alignments

- Profiles

- Sections

- Pipe networks (gravity utility networks)

- Pressure networks

- Corridors

Putting each of these in the context of the previously mentioned constructional workflow, each of these components identified in the list is a stepping stone to the next component. If we start with points (surveyed or proposed), we can ultimately create a surface from those points. If a surface is created from those points, and we change the elevation, the surface will dynamically change as well.

Sites, alignments, profiles, and sections would be in closer alignment with the framework concept where we are establishing, or designing, some basic site geometry that our BIM design model will be built on top of, acting as more of a guide.

Then we get into the heart of the design matter, where we jump into pipe networks, pressure networks, and corridors. Yes, we can design some of these BIM design components without establishing the foundational and framework elements, however, we would truly be limited in what we can do from a design standpoint.

That said, we would essentially be setting ourselves up for a lot of questions that can't be answered later on down the road. By establishing the foundational and framework elements first, we are not only setting ourselves up for a more successful design but also giving ourselves the ability to answer questions about our BIM design models at any given point in time. A thoroughly constructed BIM design model will give you the benefit and assurance in the long run that what is displayed in the drawing area can actually be constructed.

With that, we should have a fairly decent glimpse into how Civil 3D design elements function and how each is, in a way, connected to each other and can truly impact a BIM design.

Summary

In this chapter, we learned about the fundamental navigation tools and Civil 3D elements that will act as the basis for our design as we compound on this chapter going forward. Civil 3D has a very intuitive and adaptive navigation workflow to ensure the analysis and design process transitions seamlessly as we move from step to step.

With the contextual ribbon, the main commands for building upon our design automatically highlight to expedite decision making. Civil 3D provides additional methods of navigation as well as you dive further into specific workflows and begin to customize your method of working.

With a firm understanding of utilizing the navigation tools, we can effectively begin our analysis and design with the core Civil 3D elements. We learned about the different elements within the software and how they respond to different parameters and functions.

In addition to this, we learned about the Civil 3D process that links these elements together in a dynamic way to mitigate redundant work and ensure objects dependent upon each other maintain such integration through the evolution of our designs.

In the next chapter, we will learn about the details of some of these elements and dive further into their inner workings for proper and confident designs. As we explore these new elements in further detail, we will build upon the knowledge of this chapter and demonstrate their dynamic nature in incredibly fast and adaptive workflows as projects change and adjust, as is common with industry work.

Civil 3D is a program that adapts with you at your pace for maintaining focus on the design and not wasting time understanding how the program should work. Successive chapters will build upon the previous chapters, so at any point, if questions arise, feel free to return to those chapters and connect any gaps there may be.

2

Setting up the Design Environment

Now that we have begun to familiarize ourselves, and get a bit more comfortable, with the overall layout of Civil 3D, we can start to look at how to set ourselves up for successful design implementation moving forward. A critical step to any BIM design is establishing a foundation to build upon.

With that said, in this chapter, we will review how work begins within Civil 3D; we will look at how to create efficient practices from the start with file templates and settings that carry through as our drawings get more advanced and we generate more files along the way. We will also gain a bit more insight into the dynamic nature of Civil 3D and how so many of the functions within it are linked, leading to a dramatic reduction in rework and a large increase in file intelligence.

In this chapter, we will cover the following topics:

- Exploring the Toolspace within Civil 3D
- Customizing object styles and object label styles
- Expediting project work with file templates

Technical requirements

The technical requirements of this chapter are identical to the previous chapter. Please refer to the requirements listed in *Chapter 1, Introduction to Civil 3D*, for reference.

Exploring the Toolspace within Civil 3D

Going back to the *Understanding Civil 3D elements* section in *Chapter 1, Introduction to Civil 3D*, we were essentially looking at a high-level overview of the resulting major modeled components. In this section, we'll take a deeper dive into the main hub for accessing and customizing Civil 3D's objects to get a clear understanding of how all the different pieces within our drawings come together.

When we look at our **Toolspace**, you will notice four main tabs along the right side of the box (refer to *Figure 2.1*). These four tabs include **Prospector** at the top, **Settings** just below that, then **Survey**, and finally **Toolbox** at the very bottom:

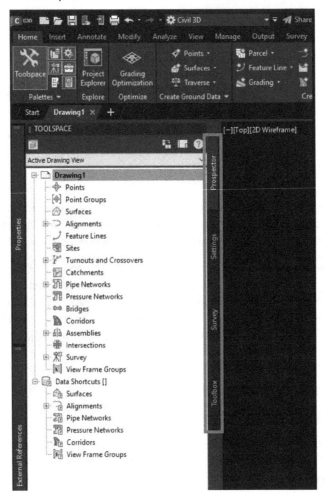

Figure 2.1 – Toolspace

When the **Toolspace** is first displayed, the **Prospector** view will automatically be displayed right away. A quick note that Civil 3D remembers which tab was displayed prior to closing your previous session, so the default tab that is displayed the next time you launch Civil 3D will be the same one displayed at the end of your previous session.

Taking a good look at what's displayed when the **Prospector** tab is selected, we can see a lot of the major modeled components that we reviewed in *Chapter 1*. The **Prospector** tab essentially allows us to create and manage individual modeled components as needed.

As modeled components are created, a + icon will appear to the left of the component category name, indicating that there are modeled components within each category. This icon allows each category of components to be expanded for further understanding of all the components contained within the drawing.

Switching over to the **Settings** tab, we can see a whole slew of additional components presented to us (refer to *Figure 2.2*). These additional components, when pieced together, are the building blocks for each of our major modeled components identified previously:

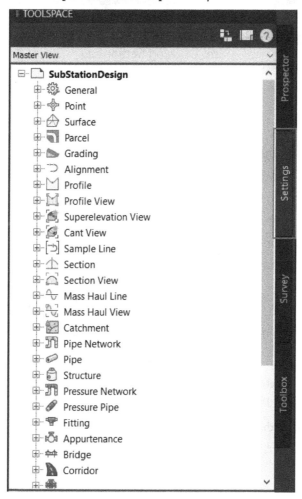

Figure 2.2 – Settings tab in Toolspace

Also within this tab, we have the ability to control all default and display settings within each specific drawing. You'll notice a similar + icon next to each component. Selecting each + icon will expand the identified category, providing a list of additional types of components, labels, and display settings that can be created and managed as needed.

Up at the top, we have the current drawing name listed. If you right-click on the drawing name here, you will be presented with some options to change various default and drawing settings associated with the current file open (refer to *Figure 2.3*):

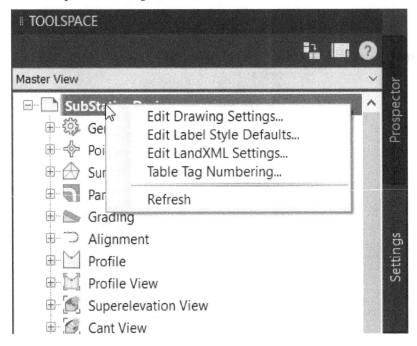

Figure 2.3 – Drawing settings right-click options

The first task, or step, I recommend everyone perform as they create a new drawing is to right-click on the drawing name in your **Settings** tab and select the **Edit Drawing Settings...** option. In the **Drawing Settings** dialog box, within the **Units and Zone** tab, you can set your drawing units, the anticipated scale of your sheets (this can be changed at any time throughout your design), along with the actual projection of your survey and/or design (refer to *Figure 2.4*).

This projection is known as the **coordinate system** and is a critical step that is best performed prior to importing any survey data or generating your design. Doing so ensures that all contents within your file are properly geolocated and in accordance with real-world coordinates:

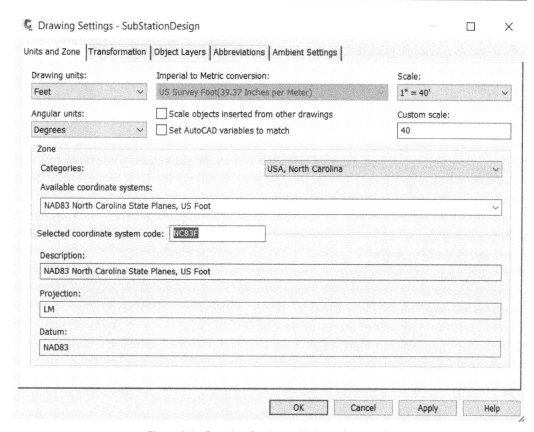

Figure 2.4 – Drawing Settings – Units and Zone tab

The next tab in the **Drawing Settings** dialog box is **Transformation** (refer to *Figure 2.5*). In this tab, you can reproject the current drawing and site contents to a new location based on the coordinate system defined in the **Units and Zone** tab we just reviewed:

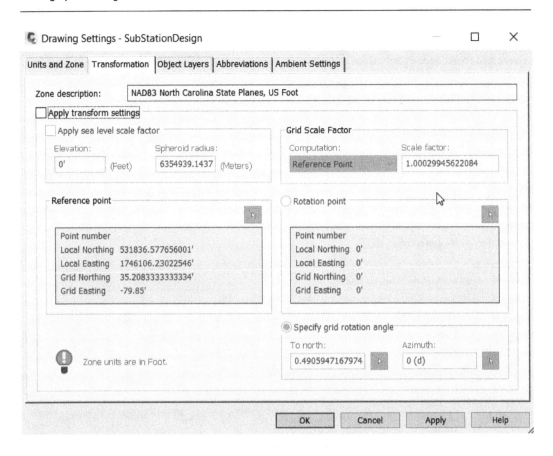

Figure 2.5 – Drawing Settings – Transformation tab

The next tab, **Object Layers**, allows end users to specify the default layer in which each modeled element is placed automatically upon initial creation (see *Figure 2.6*):

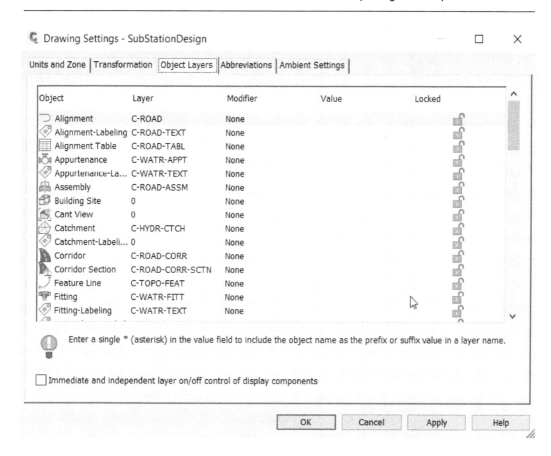

Figure 2.6 – Drawing Settings – Object Layers tab

Next up is the **Abbreviations** tab, where we specify default abbreviations that will appear within labels and reports to various modeled components and geometry within your design (refer to *Figure 2.7*):

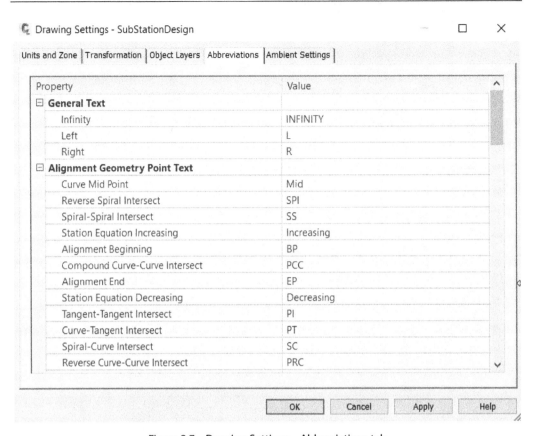

Figure 2.7 – Drawing Settings – Abbreviations tab

Last up is the **Ambient Settings** tab, which allows users to control various unit-based settings related to the current drawing (refer to *Figure 2.8*):

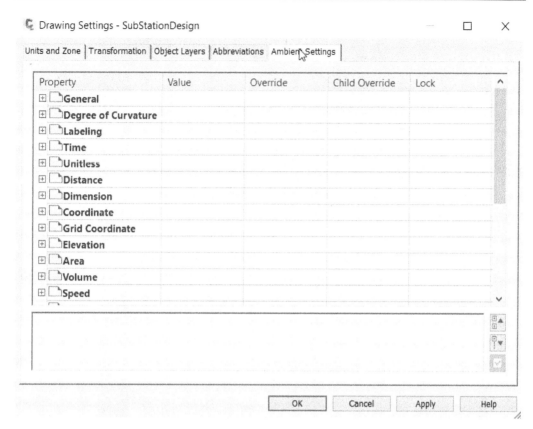

Figure 2.8 – Drawing Settings – Ambient Settings tab

When we are satisfied with our drawing settings, we can go ahead and close this dialog box, bringing us back to the **Toolspace** view. Now, we switch over to the **Survey** tab, where we can import and manage captured survey data and incorporate it into the current project (refer to *Figure 2.9*). It's important to note that survey data is managed at a project level, which would allow for the management of all files associated with that particular project:

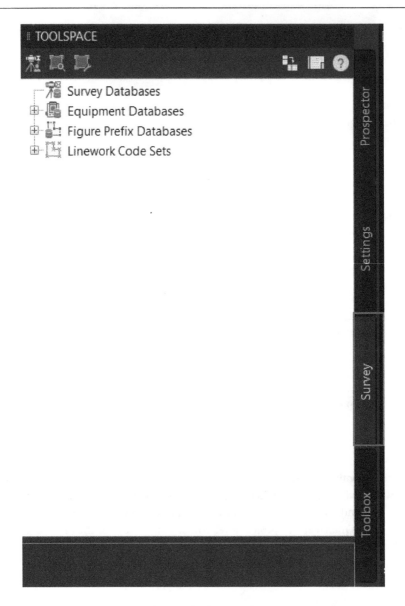

Figure 2.9 – Survey tab

The final tab in the **Toolspace** is **Toolbox**, which provides the ability to export data associated with modeled elements within your current drawing to different types of reports (see *Figure 2.10*). These reporting capabilities will come in handy in the latter phases of design:

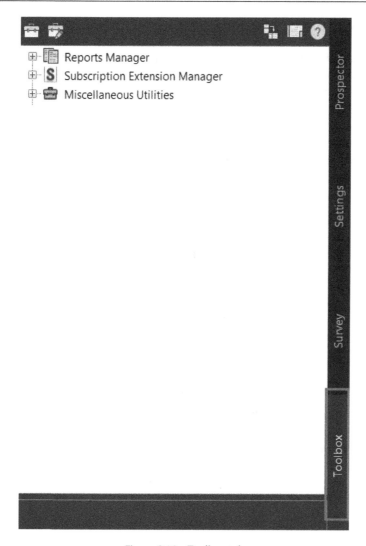

Figure 2.10 – Toolbox tab

So, that's a quick, high-level overview of the main tabs available within the **Toolspace** dialog box. We will be taking a much deeper dive into these as we begin reviewing the different types of project scenarios later, but we have briefly touched on all the portions of it. As we progress in our file creation, we will primarily be using the **Prospector** tab for design control and doing some work within the **Settings** tab.

For now, let's dive a bit deeper into the **Settings** tab to gain some insight into how we are able to create and manage the display of various modeled components, labels, and so on within our current drawing by looking into Civil 3D object styles.

Customizing object styles and object label styles

Styles are a critical component to the overall display, and final output, of your BIM design model. It's recommended to set up a drawing template that contains the majority of your already-configured display styles so that we aren't reinventing the wheel every time we start a new drawing/model. When creating new drawings, we can use this drawing template as a starting point to begin our design, ensuring that we are setting ourselves up to focus (mostly) on engineering design moving forward. It's important to note that there will almost always be a need to tweak a style here and there, or create a new custom version based on a client requirement, but going with the drawing template approach will limit the time spent on those one-offs moving forward.

When we go back to the **Settings** tab in our **Toolspace**, we can see the list of building blocks again; when pieced together, these will be used to create our major modeled design elements. When we right-click on any of these building blocks, we are presented with either two or three of the following options: **Edit Feature Settings…**, **Edit Label Style Defaults…**, and **Refresh** (see *Figure 2.11*). If a specific building block is unable to be annotated or labeled, then the **Edit Label Style Defaults…** option will not appear:

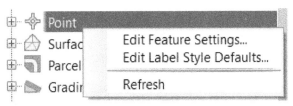

Figure 2.11 – Right-click options

By selecting the **Edit Feature Settings…** option, you will be presented with a dialog box similar to what is shown in *Figure 2.12*:

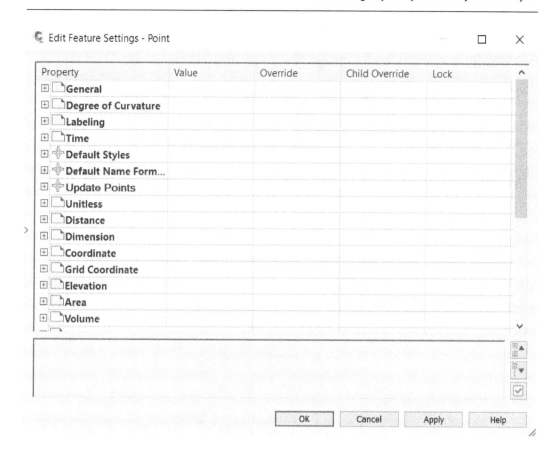

Figure 2.12– Edit Feature Settings

This will allow you to specify various settings that are particular to those individual building blocks, unlike the **Ambient Settings** options mentioned earlier, which are settings applied to an entire drawing. With that delineation, we can start to think about parent versus child settings within a drawing, whereas **Ambient Settings** is considered parent settings that are applied throughout, but a child setting can often act as an override that we are able to control as needed.

When **Edit Label Style Defaults…** is available, we are able to further define annotation and default label settings that pertain to individual building blocks of our design models, as shown in *Figure 2.13*:

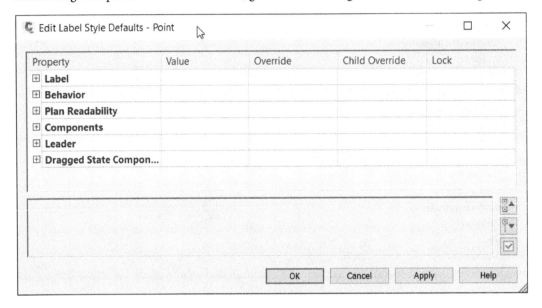

Figure 2.13 – Edit Label Style Defaults

Finally, the **Refresh** option is essentially a way to ensure that you are viewing the most up-to-date styles associated with that particular element or building block.

As we expand each element within this list, or tree, we can begin to explore all of the different settings and styles that are associated with each element (refer to *Figure 2.14*). Each element listed has a very unique purpose in building your design model out, hence the reason for the various options available as each element identified is expanded:

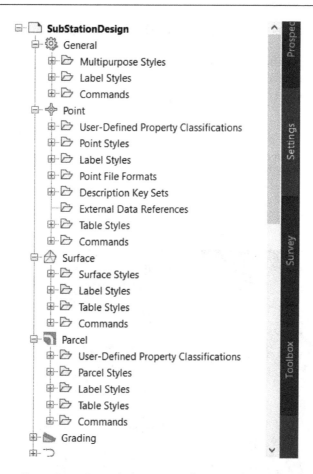

Figure 2.14 – Expanded tree view of settings and styles

As we dive more into the actual design and understand how to use all that Civil 3D has to offer us, we will then begin to explore different options for setting up our drawing templates that can be used to streamline our design moving forward and ultimately enable an efficient and consistent design throughout all our projects.

The value of establishing this work and functionality upfront can be found in the long-term benefits. As drawings progress and grow more complicated, revisions will inevitably need to be made, designs will change as more project information becomes apparent, and making changes late in a project can be detrimental to a team's success if they are not able to maintain pace with those needed changes. With these upfront templates and settings established at the start, completing project and drawing changes down the road is as easy as a few clicks.

And with that, let's finally jump into creating an actual drawing and understand how best we can utilize Autodesk's Civil 3D to generate our BIM design models and impress our clients!

Expediting project work with file templates

Creating a new drawing is the easy part; understanding the best way to create a new drawing can be a bit trickier. In this section, we'll review best practices for creating a new drawing that will drive efficiency and consistency in our designs moving forward.

As mentioned in the previous section, using drawing templates to create new drawings is ideal. Since we're just starting fresh and haven't developed our own custom drawing template as of yet, we will create a new drawing using one of the out-of-the-box templates that are made available when installing Autodesk Civil 3D.

To set Civil 3D up to use the correct drawing templates when creating new drawings, first you need to go to the top left of your Civil 3D session, open up **Menu Browser**, and select **Options**, as shown in *Figure 2.15*:

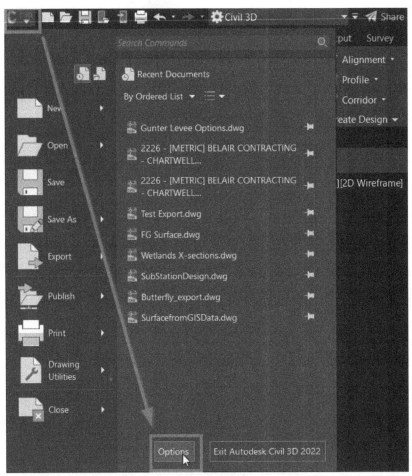

Figure 2.15 – Options selection inside of Menu Browser

After selecting **Options**, navigate to the **Files** tab and then scroll down in your view to expand the **Template Settings** section (as shown in *Figure 2.16*). This is where you can set the paths that Civil 3D will point to for recalling your library of saved drawing templates:

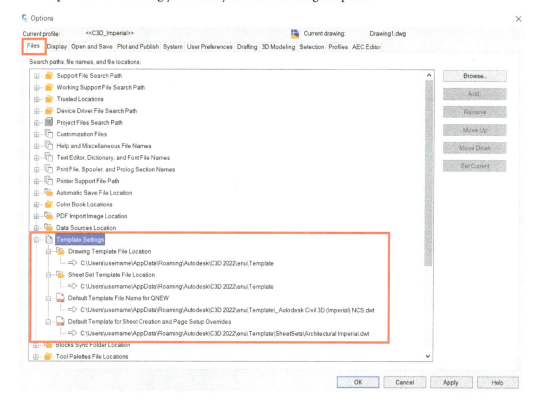

Figure 2.16 – Drawing template search paths

Any time you create a new drawing by going to **Menu Browser** and selecting **New | Drawing**, a dialog box will appear for you to select the drawing template. By default, it will recall the **Drawing Template** file location path from the **Options** dialog box as you had defined.

Any time you create a new drawing by selecting the **New Drawing** icon in the **Quick Access Toolbar** next to **Menu Browser**, a new drawing will automatically be created using the drawing template defined in the **Default Template File Name for QNEW** path within the **Options** dialog box as you had defined.

Similarly, when creating sheets using **Plan Production Tools** or **Sheet Set Manager** (these workflows will be discussed in *Chapter 15*), the paths defined in **Sheet Set Template File Location** and **Default Template for Sheet Creation and Page Setup Overrides** will be used.

With that, let's go ahead and create a new drawing, this time using **Menu Browser** and selecting **New | Drawing** (refer to *Figure 2.17*):

Figure 2.17 – Creating a new drawing using Menu Browser

After selecting **New** | **Drawing**, the **Select template** dialog box will appear. As mentioned, the default location saved in your search path will automatically be displayed. Make sure that **Drawing Template** (***.dwt**) is selected in the **Files of type** option at the bottom of the dialog box.

By changing **Files of type**, there will be different file selections displayed within the dialog box. Since we haven't created a custom template yet, let's go ahead and select the `_Autodesk Civil 3D (Imperial) NCS.dwt` drawing template and select **Open** (as shown in *Figure 2.18*):

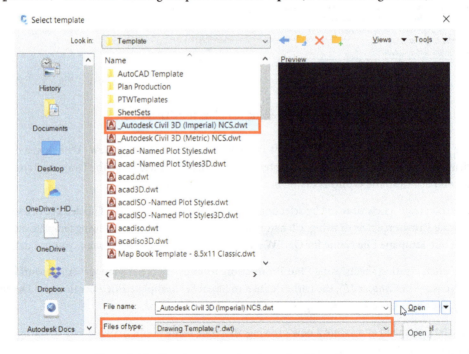

Figure 2.18 – Creating a new drawing using Menu Browser

Please note that if your design references metric units, you can select the `_Autodesk Civil 3D (Metric) NCS.dwt` drawing template to ensure that correct units are applied to your drawing.

Now we have officially created our very first drawing!

If we go back to our **Toolspace**, select the **Settings** tab, and then start expanding some of the building blocks listed, we can see that our new drawing has also been prepopulated with numerous display styles, annotation styles, table styles, default settings, and so on (refer to *Figure 2.19*):

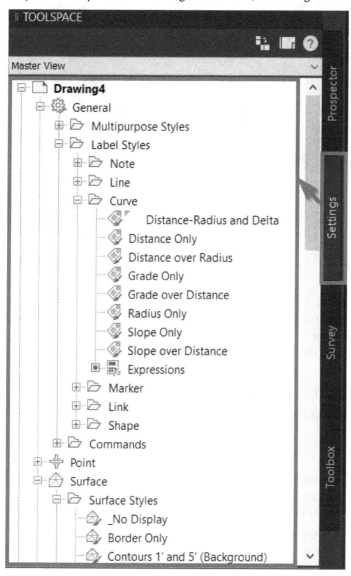

Figure 2.19 – Toolspace | Settings tab

Let's now go ahead and right-click the drawing name and select the **Edit Drawing Settings...** option to set the units and coordinate system in which our new project design resides and should be referencing. After setting the projection of the file and then clicking **OK** to close out of the **Drawing Settings** dialog box, go up to the third tier (ribbon) to locate the **Geolocation** options. Once there, select **Geolocation**, and then select **Map Aerial** to toggle on the Bing Maps-supplied aerial image (refer to *Figure 2.20*):

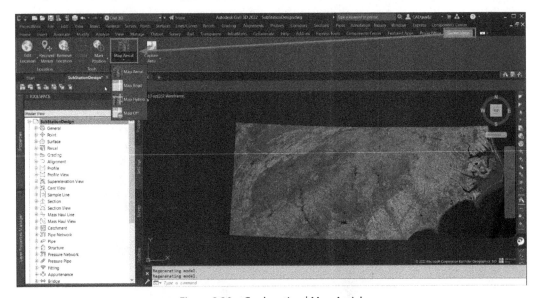

Figure 2.20 – Geolocation | Map Aerial

One thing to note is that you are only able to access this functionality when the drawing file has a coordinate system projection applied. If the drawing file does not have this defined, Civil 3D can't recognize the limits/extents, nor will it be able to detect the map that needs to be displayed in your view. This is also a great tool to validate that your survey and/or design elements are located in the right location and in accordance with real-world coordinates.

With that, we are now ready to explore options and best practices for sharing modeled elements throughout the project design, and across multiple survey and design files used to generate our full project design.

Summary

In this chapter, we continued to dive deeper into the incredible processes and functionality offered to us with Civil 3D. We now know the basics of Civil 3D and how different portions are connected and interact with each other. With that, we can optimize these connections and streamline our workflows specifically for our methods and design intent. Carrying out this up-front work ahead of time may feel a bit tedious at the start, but as you get deeper into Civil 3D and begin creating drawing after drawing and amass project after project, these templates, standardized settings, and styles will create exponential time savings for you and your team.

File templates are essential to your organization's effectiveness and ability to create professional designs and stay competitive with design and production needs and iterations. Creating an assortment of potentially required drawing templates will aid you as your group scales to different types of projects and as clients require different formats and drawing needs.

Templates, along with styles, can create a large amount of time savings not only during production but also as designs adjust. Rarely is a design perfect the first time and there are always iterations. Civil 3D allows you to keep pace with these real-world adjustments with dynamic styles and objects for linked adjustments for minimal rework when the time comes.

In the next chapter, we will learn how to expand our designs further with collaborative tools and methodologies to keep up with the size of projects that will be necessary on the market. We will dive into sharing data within Civil 3D from person to person and from drawing to drawing.

Civil 3D is not meant to house all necessary drawing elements within one drawing file, and by sharing the data and maintaining a dynamic link throughout the process, we can spread the workload out among our team for faster workflows, as well as reducing the file size. Again, if at any point in time as we progress some of this is unclear, please feel free to return to previous chapters to undertand the details and then begin adding on layers of complexity.

Sharing Data within Civil 3D

Now that we understand the benefits of configuring styles and your overall drawing settings, along with the importance of template creation, we can begin diving into best practices for managing design files and modeled objects throughout the project design development phases.

As much as we'd all like to create and pack all Civil 3D objects into one file for an entire project, this concept just isn't practical for 95% of projects you'll work on, especially if there are many objects being modeled that require dynamic interaction. On top of this, most engineers working on large projects distribute the workload among their teams to handle the size of the project, but also to expedite the creation of the project.

Managing designs and objects with Civil 3D can seem like a very daunting and complex process at first, but we'll soon realize that it's actually a very simple and straightforward process once we understand how everything gets pieced together.

Civil 3D is a powerful platform and this chapter will demonstrate how to scale your workflows with what we will refer to as data shortcuts to include more team members for larger project sizes while maintaining efficiency and keeping your design files manageable.

In this chapter, we will cover the following topics:

- Understanding file relationships
- Learning how data shortcuts work
- Creating data shortcuts within Civil 3D

Technical requirements

The technical requirements of this chapter are identical to the previous chapters. Please refer to the requirements listed in *Chapter 1, Introduction to Civil 3D*, for reference.

The exercise files for this chapter are available at `https://packt.link/UoiPn`

Understanding file relationships

When starting any new endeavor, it's best to enter with a game plan or strategy for success. Handling project designs within Civil 3D is no different; provided we take some quick measures up front to identify all the different Civil 3D modeled objects and project team members required to fulfill the project design requirements, we'll be able to ensure that we hit the ground running and maintain efficiency throughout the progression of our designs.

As we know, Civil 3D is essentially a vertical application that extends the functionality of Autodesk's AutoCAD platform, so we already have a good understanding of how external references are managed throughout the design process. While Civil 3D introduces a new level of complexity with the ability to generate civil infrastructure-related model objects, we also have the ability to data reference individual modeled objects, via data shortcuts, from one file to another as needed, eliminating the requirement to externally reference a drawing file wholesale.

I tend to use data shortcuts fairly extensively in order to manage projects and streamline the use and storage of data. They operate much like an external reference, in that when an externally referenced drawing is changed, you see the changes in the referenced drawing by reloading the reference, but you cannot modify the reference without opening the file it originated in.

Data references are similar but instead of referencing a drawing, you are referencing Civil 3D data and entities. For example, when a surface changes in file 1 and is being referenced in file 2, the user designing in file 2 can synchronize the reference and see the changes in their drawing instantly. The data that is referenced can be used for analysis, representation, calculations, and labeling, but it cannot be modified, so there is no risk of accidental modifications or deletions.

In addition to knowing when it may be appropriate, or ideal, to use external references versus using data references, it is equally important to properly organize and segregate the Civil 3D files by model and data types that will be used for your project. With that in mind, my recommendation (which is a trusted methodology across the industry) is to categorize project design files into one of the following three types of files: **model**, **reference**, or **sheet** files. This provides some additional means of properly managing files and design objects, as well as providing additional benefits of providing flexibility to allow multiple team members to update design files as needed.

Let's take a look at each of these file types.

Model files

With this structure, **model files** are intended to contain Civil 3D objects (both existing conditions and proposed design objects) that are housed in separate files dictated by the type of design objects contained within. Model files use data references to bring in the dynamic design data and use external reference files as simple overlays for the context of the entire combined drawing.

The following are a few examples of files that can be considered model files, along with some additional details of how they are intended to be set up and used.

Survey model

When the original survey file is received, it should be placed in a received folder to be left as an untouched, original file. A new **survey model** drawing will be created using the Civil 3D template and all survey information will be copied and pasted from the original file into this new file. Any surfaces or other Civil 3D entities included with the survey should be exported into a LandXML file, the `.xml` file placed into the received folder, and then imported into the new survey model file created.

The survey file should contain all existing Civil 3D objects. These can include, but are not limited to, surfaces, alignments, and pipe networks. These objects should have data shortcuts created for them and be available within the `Data Shortcuts` folder. If the surfaces, pipe networks, alignments, and so on have not been previously assigned a name, the name should reflect their status as an existing entity. For example, a surface created from data received with the survey should be named something similar to **Existing Grade**.

In the event that surface files are too large to reside within the survey model file, it is acceptable to place them in a secondary file. This file should follow a standard naming convention, such as **Topographic Survey Model**. A data shortcut should be created for the existing surface data.

Furthermore, the survey model file is considered to be an exception to the rule, in the sense that this is the only model file that will be set to a specific scale and contain all labeling and annotation that will be shown on the sheets.

Alignment model

In the **alignment model** file, alignments for roadways, dams, or linear structures will be designed. Pipe network alignments will not be placed in this file, as they are dependent on pipe network layouts. Data shortcuts should be created for these entities and the naming of proposed alignments should reflect what they represent. For example, if an alignment is set to a road named **Main Street**, the alignment should be named **Main Street**.

Grading model

In the **grading model** file, proposed surfaces will be designed. All design elements relating to the finished grading should be kept in this file, including any profiles that may be used for the purpose of creating grading objects or corridors.

Many times, grading is initiated with a corridor model, which requires a profile to be created from an alignment. The alignment should be referenced in the data shortcut created from the alignment model file, and the design profile created in the grading model file. In turn, there should be a data shortcut created for this design profile for use in other model files, as needed.

The naming of the proposed surface(s) should reflect what the surface is for. For example, if the surface is for the entire site grading of a subdivision, it should be named **Site Grading Subdivision**. If it is for a road grading of Highway 123, it should be named **Highway 123 Roadway**. Data shortcuts should be created for these surface objects to help maintain flexibility and control of specific sections of a design.

Utility model

In the **utility model** file, pipe networks will be designed. All design elements relating to the pipe network will be kept in this file, including any profiles that may be used for designing the utility system. Depending on the complexity of the site, it may be necessary to create several utility model files, one for each different type of piping network, such as storm, sanitary, water, and gas. Each pipe network should be named logically, such as **Sanitary Sewer** or **Water Main**. Data shortcuts should be created for these structure and pipe objects separately again for finer control as modifications need to be made.

Reference files

Next up are **reference files**, which are intended to contain/represent 2D geometry and static elements and annotation. Reference files would include content such as surveyed planimetrics, civil site plan geometry, and erosion control BMPs.

The following are a few examples of files that can be considered reference files, along with some additional details of how they are intended to be set up and used.

Site plan reference file

In the **site plan reference file**, there should be points created for coordinate table generation (commonly used for site staking plans), if required. These points should be placed in a point group named appropriately to reflect what they represent. For example, if points are placed to designate a parking lot layout, the point group should be named `Parking Lot Layout Points`.

If alignment tables need to be created, the alignment(s) will be data referenced into the site plan reference file. Tags will then be assigned, and tables generated, from the alignment as required.

Generally, topographic data is not shown on site layout plans, but if required, it will be brought into the site plan model file via a data reference from the survey and/or the grading models.

Grading reference file

In the **grading plan reference files**, there should be the existing and proposed surfaces data referenced in from the survey and grading model files. All surface labels and any needed tables will be created in this file at the scale designated.

Utility plan reference file

In the **utility plan reference files**, pipe networks, alignments, and surfaces will be data referenced into this model from the alignment model files. Labels, tables, and any other necessary annotation will be placed in this file at the scale designated.

Profile reference file

In the **profile reference file**, all alignments, design profiles, and surfaces will be data referenced into this reference from the model files. This data will be used to produce profile views within this file. Labels, tables, and any other necessary annotation will be placed in this file at the scale designated.

Section reference file

In the **section reference file**, all needed alignments, design profiles, pipe networks, and surfaces should be created with data references. This data will be used to produce section views within this file. Labels, tables, and any other necessary annotations will be placed in this file at the scale designated.

Sheet files

Lastly, **sheet files** are the final product of the project. The construction of the sheet files will consist of externally referencing both the reference and model files, along with sheet borders, general notes, north arrows, and any additional sheet-specific annotation.

Following a project structure like this has many benefits, ranging from overall design file optimization to promoting work-sharing opportunities across your project team(s). Regardless of the size of your project and team(s) involved, incorporating this methodology can only increase efficiency in your design process. This breakdown of work ensures that project team(s) are able to stay focused on all of the various tasks required by allowing multiple team members to work on different files and design models simultaneously.

Now, let's dive into the details of how data shortcuts work for such a flexible and dynamic work structure.

Learning how data shortcuts work

As mentioned, data shortcuts allow project teams to share and reference specific Civil 3D modeled objects from one drawing to another. This functionality allows project teams to maintain high levels of efficiency by not only referencing just the Civil 3D objects needed at that time but also limiting the amount of content and data residing in your current drawing.

Data shortcuts work by decentralizing a project's content, allowing each large entity (i.e., a surface, pipe networks, and corridors) to exist in its own separate drawing. By creating data shortcuts, you tell each drawing which drawings are related to it or are in the same project, and then are able to sample that data within the current drawing.

The alternative to this is housing each large object in the same file, dramatically increasing the file size and similarly reducing the maneuverability within that drawing. This separation of objects is not only beneficial for effective space allocation within files but also allows for more collaboration on files, having team members perform actions simultaneously for faster project completion.

Often, especially with earlier releases with Civil 3D, the size of the file can increase exponentially, commonly referred to as **file bloat**, causing significant delays and refresh rates while performing even the simplest of commands. It is important to note that, starting with Civil 3D release 2024, Autodesk made significant improvements to overall performance and file optimization to alleviate some of these issues experienced with earlier releases.

Getting your Civil 3D environment ready to use data shortcuts is best configured at the beginning of a project, similar to setting up your drawing settings and drawing templates, as discussed in *Chapter 2, Setting up the Design Environment*. There are essentially two functions needed for the implementation of data shortcuts: creation and referencing. The data shortcuts must be *created* within the file that the data exists in, and then they must be *referenced* in the file where the data needs to be shown or used.

The first step in getting your Civil 3D environment ready to use data shortcuts is to pull up the **Toolspace** again and go to the **Prospector** tab, as shown in *Figure 3.1*:

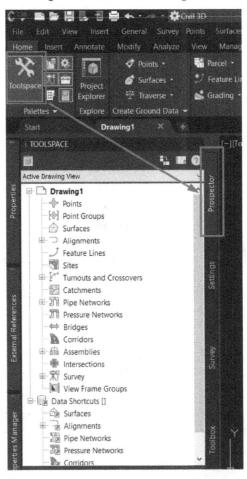

Figure 3.1 – Toolspace Prospector tab

Next, we'll scroll down to the area in **Prospector** that is labeled **Data Shortcuts []**, right-click on the text, and select **Set Working Folder…**, as shown in *Figure 3.2*:

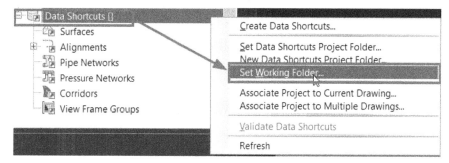

Figure 3.2 – Set Working Folder… selection

In the **Set Working Folder** dialog box, we must first identify and then navigate to the location in which we would like to store our project's data shortcuts (refer to *Figure 3.3*):

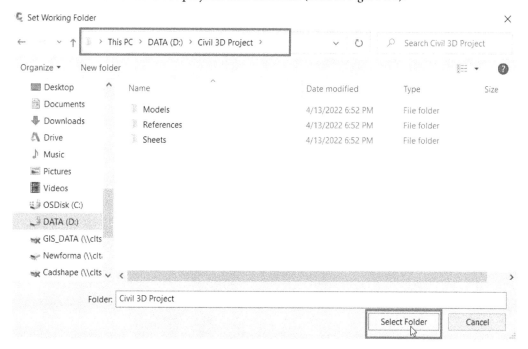

Figure 3.3 – Set Working Folder dialog box

My recommendation would be to store these at the root of your **Project Design** folder. We will then assign the data shortcuts project folder a unique naming convention to ensure that we can quickly identify it later on, if needed, and keep it consistent when we associate them with the corresponding project design files.

Once selected, we now need to go back to **Prospector**, right-click on **Data Shortcuts** [] again, and now select **New Data Shortcuts Project Folder…**, as shown in *Figure 3.4*:

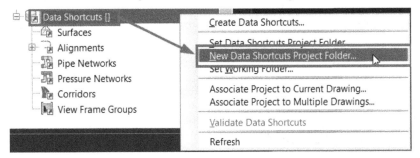

Figure 3.4 – New Data Shortcuts Project Folder…

A recommended practice is to give the data shortcuts project folder a standard naming convention consisting of the Autodesk program, version, project number, and, finally, the Data_Shortcuts annotation to alert others within the project of the purpose of the folder and files contained within and it is not just a working or temporary folder. For example, a project being produced with Civil 3D version 2024 with a project number of 123456 would have a data shortcut project folder named C3D_2024_123456_Data_Shortcuts, as shown in the **New Data Shortcuts Folder** dialog box in *Figure 3.5*:

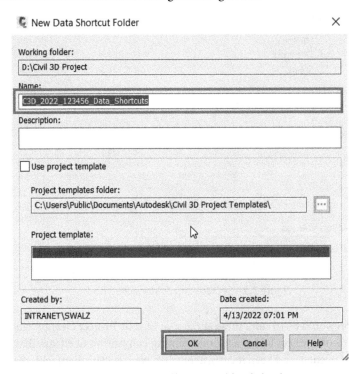

Figure 3.5 – New Data Shortcut Folder dialog box

Immediately after your new data shortcuts folder has been created, you'll notice that the path and name of the data shortcuts project are now listed in brackets next to **Data Shortcuts** in **Prospector**. Your drawings, which need to have access to the project data, will now need to be associated with that project folder. This is done by right-clicking on the **Data Shortcuts** section of **Prospector** one final time and selecting **Associate Project to Current Drawing…**, as shown in *Figure 3.6*:

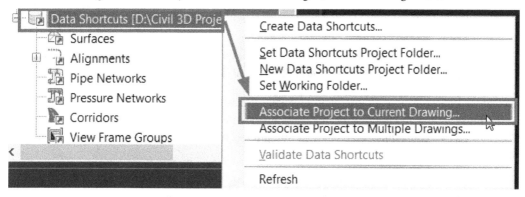

Figure 3.6 – Associate Project to Current Drawing…

Note

In *Figure 3.6*, you'll notice there is an **Associate Project to Multiple Drawings…** option. Selecting this option allows you to create your drawing first and then apply the project association process to all files in a given folder or subfolder. This practice is not ideal, but it's important to note that it is an available option if these steps are missed during the initial project setup.

Once the **Associate Project to Current Drawing** dialog box appears, simply select the data shortcuts project you just created (refer to *Figure 3.7*) and you'll be all set to move forward:

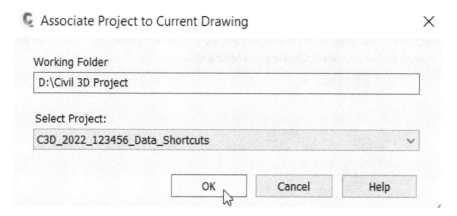

Figure 3.7 – Associate Project to Current Drawing dialog box

It's important to note that each file where Civil 3D modeled objects are being developed will need to be associated with the corresponding project. Associations ensure that the drawing files retain the location where the project data should be pulled from. Once the drawing is properly associated with the project, you may also notice that the drawing title header at the very top of your Civil 3D session will also reflect which project it is assigned to.

Now that we've learned about how data shortcuts work within a larger project, we can take the next step to collaborate further with our designs.

Creating data shortcuts within Civil 3D

Now that we know how to configure data shortcuts and set our project design up from the get-go, let's dive into creating and linking specific Civil 3D modeled objects from one drawing to another. We'll also look into some quick management tips and recommended practices for sharing Civil 3D modeled objects.

As we create Civil 3D objects (i.e., surfaces, alignments, gravity/pressure networks, corridors, etc.), we can begin to think through what additional types of model and reference files will be needed for the detailed design in preparation for sheet file creation.

Thinking through what that may look like, let's use the example of an existing surface created in our survey file. What other types of design files (model and/or reference) will need to reference and display the existing surface for detailed design purposes? A quick analysis would determine that the existing surface would need to be present in the proposed grading model, utility, profile, and section model files.

In some of these files, other Civil 3D objects will need to be present as well; for example, the existing utility Civil 3D objects would need to be present in the proposed utility model file to be able to tie proposed networks into the existing one. In that same sense, existing utility Civil 3D objects would not be required to be present in the proposed grading model.

So, let's run through the process of creating data shortcuts. As shown in *Figure 3.8*, we can see that within our existing **survey model** file, we have an existing surface created. To allow for this type of Civil 3D object to be data referenced into other files, we'll need to right-click on the data shortcuts project we created earlier and select **Create Data Shortcuts…**.

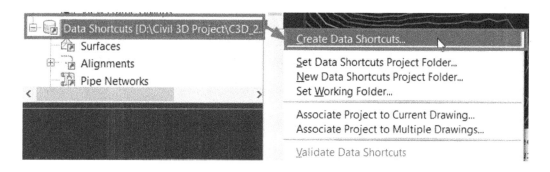

Figure 3.8 – Create Data Shortcuts… on a surface model

After selecting the **Create Data Shortcuts…** option, you'll be presented with a **Create Data Shortcuts** dialog box (as shown in *Figure 3.9*). In this dialog box, you'll have the opportunity to add Civil 3D modeled objects from the current drawing to your data shortcuts project.

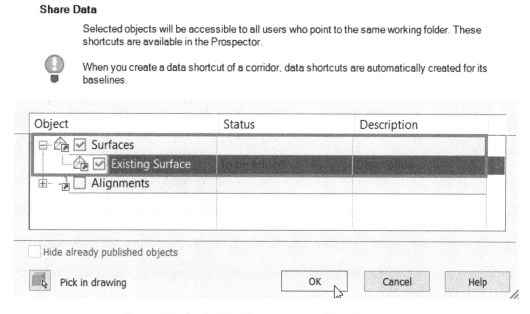

Figure 3.9 – Create Data Shortcuts page of a surface model

In this case, we have the opportunity to create data references for both **Surfaces** and **Alignments**. If we select the checkbox next to **Surfaces** and then select the **OK** button at the bottom of the dialog box, we can then share these particular Civil 3D modeled objects across multiple drawings, provided they are associated with the same data shortcut project.

After selecting **OK**, you'll notice that **Existing Surface** has been added to the data shortcuts project, as shown in *Figure 3.10*. As we create new files or open existing files that are associated with the same data shortcuts project, we'll be able to data reference in the existing surface to further our design:

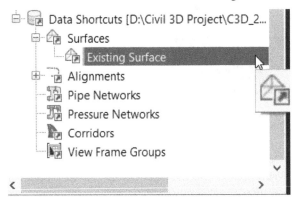

Figure 3.10 – Surface model data shortcut

To further enhance our data referencing project management, we also have the ability to create folders to organize our Civil 3D objects. This definitely comes in handy as project designs become more complex. Thinking simplistically, we can create `Existing` and `Proposed` folders within each Civil 3D modeled object. More complex projects can expand upon this by creating site- or area-specific folders, main alignment or roadway folders, utility use folders, and so on.

One thing to note is that folder creation can occur at the current drawing level, as well as the **Data Shortcuts** level. If we were to create this folder structure at the drawing level, we would need to right-click on the Civil 3D modeled object, in this case, **Surfaces**, and select the **Create Folder** option (refer to *Figure 3.11*):

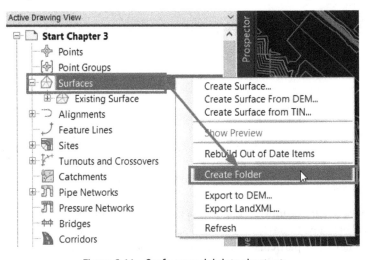

Figure 3.11 – Surface model data shortcut

In the **Create Folder** dialog box that appears, we can type the name `Existing` and click **OK** (refer to *Figure 3.12*):

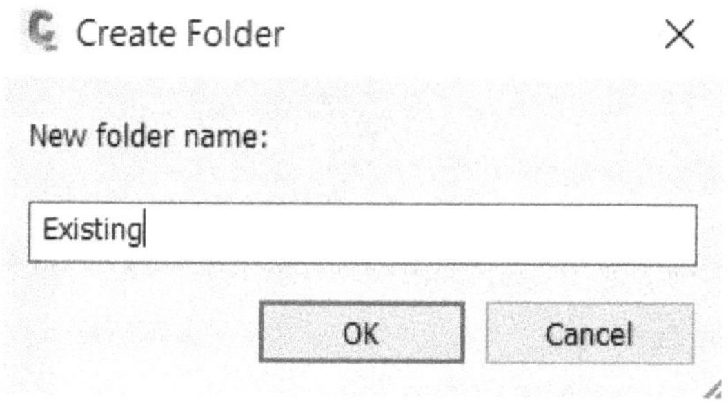

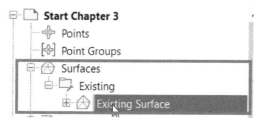

Figure 3.12 – Create Folder

Once created, we can then select **Existing Surface**, which has already been created, and then drag and drop it into the **Existing** folder, as shown in *Figure 3.13*:

Figure 3.13 – Reorganizing Existing Surface at the drawing level

If we now go back to our data shortcuts project, shown in *Figure 3.14*, right-click on **Existing Surface** that we just added, and select **Remove**, the surface will then be removed from the data shortcuts project.

When we recreate this data reference moving forward (going through steps outlined earlier in this section and shown in *Figures 3.8–3.10*), we can now see that an **Existing** folder has been created automatically, and **Existing Surface** resides within, as shown in *Figure 3.14*:

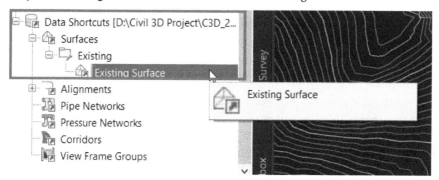

Figure 3.14 – Add Existing Surface with the drawing level folder structure

Alternatively, if we decide to manage folders at the data shortcuts level, and not within each individual drawing, we have the opportunity to do so by right-clicking on the Civil 3D modeled objects in our data shortcuts project and selecting the **Create Folder** option here, as shown in *Figure 3.15*:

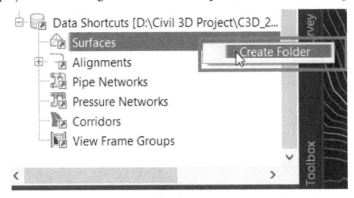

Figure 3.15 – Creating a folder in the data shortcuts project

After we create the folder and create the data reference again, you'll notice that **Existing Surface** is added, but uncategorized, as shown in *Figure 3.16*. We'll then need to select **Existing Surface** in the data shortcuts project and drag and drop it into the **Existing** folder:

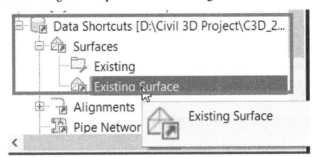

Figure 3.16 – Reorganizing Existing Surface in the data shortcuts project

Out of the two options, my personal preference is to create folders and manage them at the data shortcuts project level. With this option, it's a one-and-done kind of deal. Although Civil 3D objects are not automatically categorized as they're created, it really shines a light on proper data shortcuts management and driving consistency across your entire project design.

As we begin to expand our teams involved in these projects and introduce more files and data management requirements, data shortcut project management has the potential to become a full-time job if not handled properly. With that, it is ideal to handle folder structures and proper organization of Civil 3D modeled objects in one location (the data shortcuts project), rather than all individual files that are created for the project that contain Civil 3D modeled objects.

Summary

As covered in this chapter, there are many things to consider when setting up your project to ensure that it can be properly managed throughout the design phase. Depending on the size of your design team and project scope, proper file organization and referencing procedures/workflows can be critical to maintaining high levels of design and collaboration efficiency.

In this chapter, we learned how Civil 3D allows for more efficient file management for not only saving space and maintaining maneuverability within drawings but also allowing teams to collaborate easier between phases of a project, as well as working simultaneously together. This can be crucial in today's real world of increasingly complex design projects.

The majority of Civil 3D design projects today will require creating and managing multiple data shortcuts with multiple external references, and this up-front management strategy can save hours and days for your team as work is coming in and the project clock starts ticking.

As we jump into the next several chapters, we'll begin to realize how data referencing can be utilized to our advantage. In the next chapter, we will begin to get stuck into Civil 3D for more hands-on work past the foundational understanding of how the strategies of this program work and why they do in this way.

We'll also begin to pull all of the design pieces that we've learned in the first three chapters together and understand how all of these settings, configurations, and workflows can be applied within an actual project design setting.

Part 2: Designing and Modeling with Civil 3D from Scratch

In this next part, we will dive deeper into the detailed components of Civil 3D to understand how the individual control and analysis of your existing conditions turn into an exceptional design.

The following chapters are included in this section:

- *Chapter 4, Configuring Survey Data with Civil 3D*

- *Chapter 5, Leveraging Points, Lines, and Curves*

- *Chapter 6, Surfaces - The First Foundational Component to Designs within Civil 3D*

- *Chapter 7, Alignments - The Second Foundational Component to Designs within Civil 3D*

- *Chapter 8, Profiles - The Third Foundational Component to Designs within Civil 3D*

4

Configuring Survey Data with Civil 3D

As we have begun to develop a fundamental understanding of how Civil 3D operates and have reviewed in a bit of detail many of the foundational settings, configurations, and workflows necessary to begin a design project within the Civil 3D environment, we can now start to put these skills to the test and into practice.

Typical civil design projects tend to begin with the development of a Survey Model representing your site's existing conditions. In this chapter, we will demonstrate the start of bringing in existing survey data with proper formatting and how to establish an accurate depiction of the yet-to-be-designed project site.

That said, key topics that we will cover in this chapter include the following:

- Survey setup
- Introduction to the Survey Toolspace
- Existing conditions display settings
- Survey workflow overview
- Analyzing your existing conditions

Additionally, starting with this chapter, we will begin to use a real dataset to get some hands-on training as we learn the different tools that Civil 3D has to offer us, and how we can leverage them for real-world design applications. We will begin by using the dataset found in our `Practical Autodesk Civil 3D 2024\Chapter 4` subfolder, starting with the file titled `Survey Model_Start.dwg`.

Technical requirements

It's important to note Autodesk's Civil 3D can oftentimes be very taxing on your computer. There is a lot of processing that goes on with modeled design elements, even in the background, that enables the dynamic (connected) capabilities to occur throughout the **Building Information Modeling (BIM)** design lifecycle.

In turn, there are many technical requirements that need to be considered to allow Autodesk's Civil 3D to operate at its full potential. We'll review the minimum requirements that Autodesk recommends, with a few of my suggestions added to increase efficiency and speed throughout the BIM design process:

- **Operating system**: 64-bit Microsoft Windows 10
- **Processor**: 3+ GHz
- **Memory**: 16 GB RAM (I suggest going with either 64 GB or 128 GB)
- **Graphics card**: 4 GB (I suggest going with 8+ GB)
- **Display resolution**: 1980 x 1080 with true color
- **Disk space**: 16 GB
- **Pointing device**: MS-Mouse compliant

The exercise files for this chapter are available at `https://packt.link/UoiPn`

Survey setup

Setting up your Survey Model properly from the get-go can be compared to setting the foundation for a house. All design elements will essentially be built on top of—and will need to reference in—various methods, the Survey Model depicting the existing built environment.

In this section, we'll run through the basics of Civil 3D survey tools and get a clearer understanding of how we can leverage each to configure or set up our Survey Models. This section will act as our inspection phase of the survey to investigate the conditions of the drawing, including ensuring the geolocation has been properly set, determining how the existing conditions were modeled, and checking the entire drawing for miscellaneous drawing elements to ensure the file is neat and ready to be used as the foundation of our design.

Let's go ahead and open up our `Survey Model_Start.dwg` file. Once it's open, you should see your overall existing conditions, or Survey Model, being displayed, as shown in *Figure 4.1*:

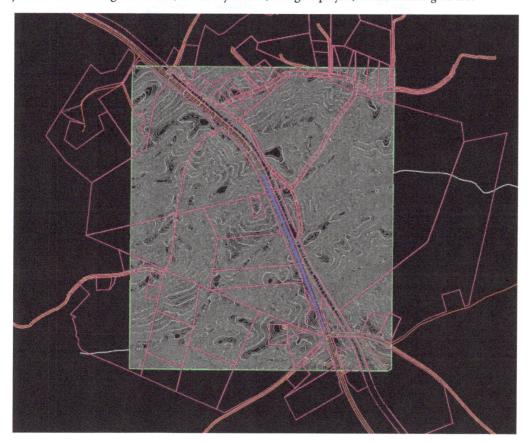

Figure 4.1 – Survey Model

If we go ahead and launch our Toolspace (accessed through our **Home** ribbon) and select our **Prospector** tab within the Toolspace, we will begin to see some Civil 3D modeled objects already present within our file, as shown in *Figure 4.2*:

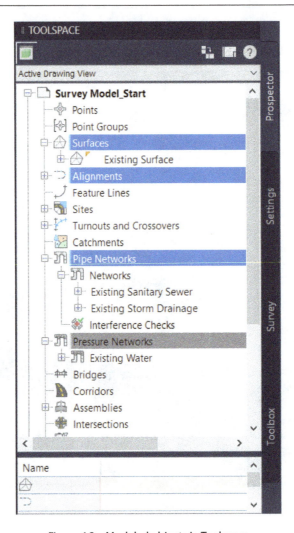

Figure 4.2 – Modeled objects in Toolspace

These Civil 3D objects consist of **Surfaces**, **Alignments**, (gravity) **Pipe Networks**, and **Pressure Networks**). Each of these Civil 3D objects within our prospector can be expanded by clicking on the + icon next to each object for a more detailed breakdown of the elements that define each object.

The next step I recommend taking before going any further is ensuring that the coordinate system projection has been properly configured for the Survey Model. As discussed in *Chapter 2*, *Setting up the Design Environment and Setting Your Next Project Up for Success*, this is a critical step that should be performed to provide you with the assurance that your design is projected into real-world coordinates.

That said, let's go ahead and select the **Settings** tab on our Toolspace. Right-click on the drawing name listed at the top and select **Edit Drawing Settings**.

Once it's selected, we will be presented with our **Drawing Settings** dialog box where we can set (or confirm, in this case) that our Survey Model has been assigned a coordinate system, as shown in *Figure 4.3*:

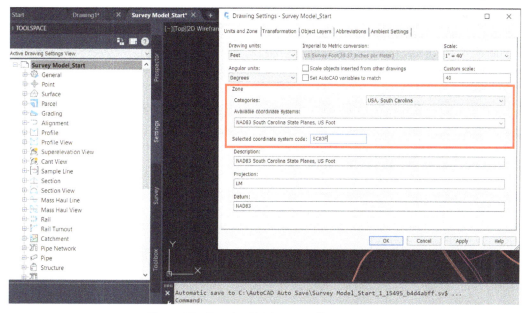

Figure 4.3 – Drawing Settings coordinate system

Now that the coordinate system has been confirmed, let's click on **OK** at the bottom of the **Drawing Settings** dialog box and go back up to our ribbon toward the top.

Let's select the **Geolocation** ribbon, go to our **Online Map** section, and change our **Map Off** selection to **Map Aerial**. Once this has been changed, an aerial image of our site and surrounding areas will be displayed as a background behind our Survey Model, as shown in *Figure 4.4*:

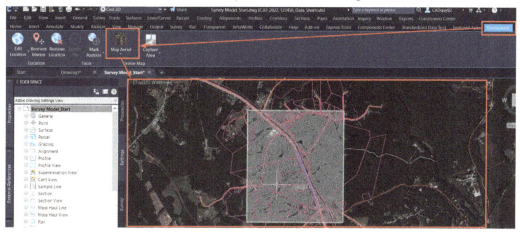

Figure 4.4 – Map Aerial from the Geolocation ribbon

It's important to note that if the coordinate system were set up to a different projection and we then proceeded with toggling on the **Map Aerial** selection, we would be presented with a completely different map associated with that projection and the map would no longer line up with our site.

That said, a quick visual inspection can confirm that the coordinate system that the file has been associated with is in fact the true projection of the site, essentially meaning that we're good to go and ready to start proceeding with continued Survey Model setup in preparation of our design for this project.

Next, let's investigate the elements contained within the drawing. At first glance of the Toolspace, we saw Civil 3D recognized the modeled components of a surface, pipe networks, a pressure network, and some alignments. Let's run through a quick exercise to inspect these elements further.

Survey files can come in many shapes and forms. The exercise file we will be using contains modeled Civil 3D components that are dynamic, just as Civil 3D would prefer, and would be able to fully leverage further into design. However, files do not always come like this. A majority of survey files received come as 2D linework drawings with flat contours and symbols for utilities and more.

Civil 3D is, to be put frankly, a three-dimensional program, and many who utilize it prefer it to remain as a 2D program. So, when receiving files, do not be surprised when you see elements flattened with their elevation set to 0'. What can also happen is some elements have their elevation set to 0' while others have been missed, so their linework may be at 0' as well as 800', or whatever the local elevation of the region is. This is why it is so vital to inspect your survey drawings thoroughly and understand how they were generated and what you need to do to be successful with them.

Now that we have inspected the modeled elements and ensured the geolocation is accurate, let's walk through some simple housekeeping of the drawing to ensure it is ready to be used for our design.

To verify that all objects in our Survey Model are geolocated and in close proximity of our site, we can perform a simple Zoom Extents of our file to quickly see if any objects fall outside of our site, or are significantly far away. If objects are far away from our site, we can select those objects using the simple Window Selection, right-click with our mouse, and select **Properties** so that we can identify which types of objects they are, along with any additional information associated with them.

Occasionally, surveyors and engineers will place various symbology, blocks, annotation, and so on on the side so that they can quickly access these objects to populate the Survey Model, and then forget to remove them later on from the file before handing off to the design firm. Although these remnants typically are not a cause for concern, it is recommended to remove these objects if they are not contributing or adding any value to our Survey Model or design moving forward. This way we are keeping things clean from the start.

> **Quick exercise with zoom extents**
>
> Do an exercise by turning on **Properties** to the right of the screen and select elements in the drawing, noting in **Properties** what you have selected, checking for miscellaneous linework, and so on.

To conclude this section, the survey files you work with are again foundational to the design moving forward. The more accurate the survey file is, the more accurate and confident you can be in your design. It is vital to always inspect your survey files, whether they come from your in-house surveying department or an outside surveying firm.

Each group conducts itself differently and uses different nomenclature and techniques for generating these legally binding documents for use in projects. You may not always receive the most up-to-date Civil 3D drawing with modeled components and existing pipe networks modeled. Most of the time, you can expect to receive a flat 2D drawing with contours and basic notations for utilities and materials.

Running through the files you receive and fully inspecting all the conditions listed previously can save you time and headaches before you begin to design and realize down the road that something is off that wasn't seen earlier. With this first section complete, let's move on to further understanding Civil 3D's capabilities with surveying.

Introduction to the Survey Toolspace

Oftentimes, the **Survey** tab within our Toolspace can help us better manage survey data and settings within our project files. It's important to note that survey data and settings being managed in this space are applied to our entire Civil 3D project and are not drawing-specific.

Also, it's very important to note that although this functionality can come in handy from a surveyor standpoint, it is not typically utilized from a design standpoint. Nevertheless, let's jump into how to access the Survey Toolspace and understand some of the capabilities contained within.

To access it, we'll need to go back to our Toolspace on the **Home** ribbon and select our **Survey** tab along the right side of our Toolspace, as shown in *Figure 4.5*:

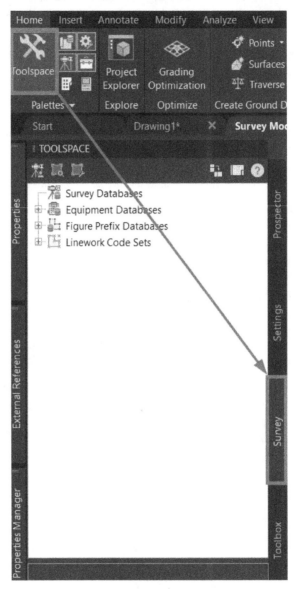

Figure 4.5 – Survey tab in Toolspace

Once presented with the Survey Toolspace, we now have access to configure **Survey Databases**, **Equipment Databases**, **Figure Prefix Databases**, and **Linework Code Sets**, all by simply right-clicking on any of these functions listed out and selecting **New**.

Here's a quick overview of what each of these selections allows us to do:

- **Survey Databases**: Allows the configuration of survey-related networks, including control points, non-control points, known directions, observations, setups, and traverse definitions, along with figures and survey points

- **Equipment Databases**: Allows the input of specifications related to the equipment used to survey your site

- **Figure Prefix Databases**: Provides the ability to set the layer that figures are placed on within our Survey Model

- **Linework Code Sets**: Allows the mapping of field codes utilized in the data collector used by the survey field crew

Through these selections, surveyors are able to make connections to data collectors used in the field to import and convert data into real geometry within Civil 3D. As more data is collected, surveyors can continue to update the geometry by either manually updating or hosting a dynamic link to the project that will push updates automatically into your files.

From a designer perspective, we typically get involved after all of the survey data has been converted into geometry where the surveyor will then hand off the Survey Model (in our case, this will be our `Survey Model_Start.dwg` file) to the engineer/designer where we can then clean it up to conform to any display standards required for the project.

In this section, we briefly dived into the detailed functionality of the Survey Toolspace and were able to get more of an understanding of all the capabilities Civil 3D can offer the surveying industry. Since this book is intended to focus on design principles and workflows, we will not need to explore them in more detail as our intention for design differs.

For now, we know where to access the tools if the need may arise and we understand how we can affect the survey if needed, but the best practice for working with survey files is to understand them and take stock of what was delivered and not to modify the drawing at all lest we risk modifying the existing conditions incorrectly.

Next, we will explore the settings that control the appearance and characteristics of our existing conditions. With an understanding of what a survey file contains, as well as all that is entailed when working with a survey file, we can apply what we learned to adjust the file visually for a better understanding of our design strategy without modifying the actual survey file that was delivered to us as the designer.

Existing conditions display settings

Now that we have received the `Survey Model_Start.dwg` file from the surveyor, we can begin looking into ways to clean it up to meet our project design and display requirements. It is important for us to understand what is advised to do and not to do when working with survey files.

Earlier, we inspected all the aspects of our drawing to ensure its integrity, but now we will be exploring how we can modify its appearance to better understand our existing conditions without risking the modification of the actual drawing itself.

If you have not done so already, let's go ahead and open the `Survey Model_Start.dwg` file contained within our `Practical Autodesk Civil 3D 2024\Chapter 4` dataset folder.

Once opened, we'll then click on the **Manage** ribbon and select our **Import** icon in the **Styles** section on the ribbon, as shown in *Figure 4.6*:

Figure 4.6 – Import styles

We'll then be prompted with a dialog box to search for the template file that we'd like to import. At this point, we're going to want to locate and import a client-based Civil 3D template, or a company Civil 3D template, if available.

If not, we can always start off with the standard Imperial template that comes with standard Civil 3D installations to use as a foundation for us to begin developing our own company Civil 3D template moving forward.

With that, let's go ahead and navigate to our local directory at `C:\Users\replacewithusername\AppData\Roaming\Autodesk\C3D 2024\enu\Template` and select our `_Autodesk Civil 3D (Imperial) NCS.dwt` file to import basic styles into our drawing, as shown in *Figure 4.7*:

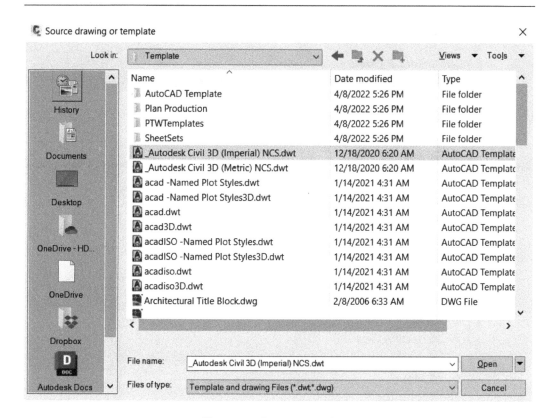

Figure 4.7 – Drawing templates

We'll next be prompted to identify which label, table, and display styles from the Civil 3D template we'd like to import into our current drawing. Feel free to expand any of the sections to get a better understanding of what will be included in the importing process.

However, since we're essentially starting from scratch, let's just go ahead and check the box at the top that selects all styles to import (as shown in *Figure 4.8*) and click on **OK**:

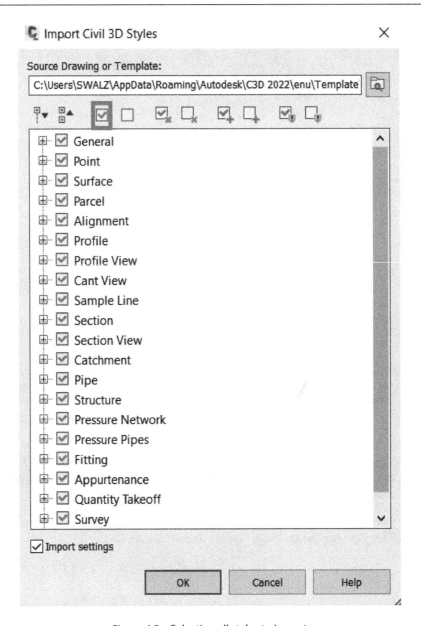

Figure 4.8 – Selecting all styles to import

Now that we have a great foundation to start with, we can now begin standardizing and assigning display styles to our modeled survey elements. We'll now go back to our **Prospector** tab in our Toolspace, expand the **Surfaces** category, right-click on **Existing Surface**, and select the **Surface Properties…** option, as shown in *Figure 4.9*:

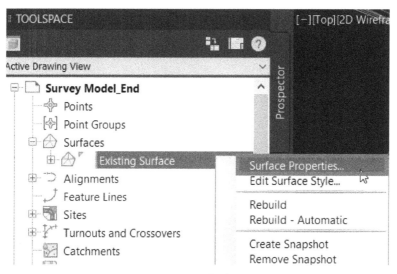

Figure 4.9 – Accessing Surface Properties

We will be presented with our **Surface Properties** dialog box, shown in *Figure 4.10*:

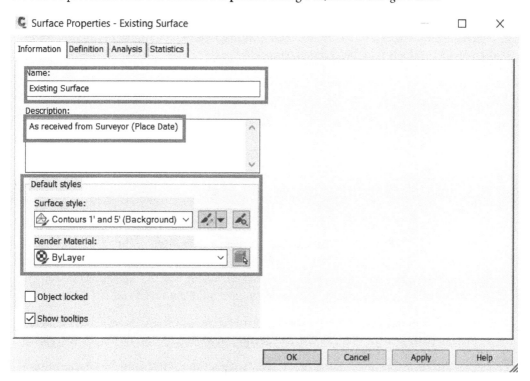

Figure 4.10 – Surface Properties dialog box

In the **Surface Properties** dialog box, we can modify the name, description, and default styles associated with the surface. The **Description** input is often overlooked, but it's definitely good practice to provide some level of description as it relates to how and/or when a modeled object was constructed.

You'll notice the **Default Styles** section has two options for display: **Surface style** and **Render Material**. The **Surface style** option will be the most common style used for plan view purposes and is applied whenever the Wireframe visual styles are applied.

The **Render Material** option will only be displayed when in a rendered state or when a realistic visual style is applied. For now, let's keep the **Surface style** option selected as **Contours 1' and 5' (Background)** and the **Render Material** option as **ByLayer**.

To dig a little deeper into the display style settings, let's go ahead and select the down arrow icon directly next to the **Surface style** selection. You'll notice a drop-down list of options made available where we can do the following:

- **Create New**: Allows you to create a new display style from scratch and refine the settings within

- **Copy Current Selection**: Allows you to make a copy of the current selection and refine the settings within the new version

- **Edit Current Selection**: Allows you to refine the settings within the currently selected display style

- **Pick from Drawing**: Allows you to select another display style of a similar object within your current file and apply it to the current object

Let's go ahead and select the **Edit Current Selection** option, as shown in *Figure 4.11*:

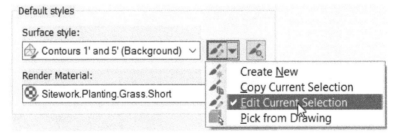

Figure 4.11 – Edit Current Selection: surface style

Once selected, we'll then be presented with a **Surface Style** dialog box that allows us to refine the settings of the currently selected surface display style—in this case, **Contours 2 and 10 (Background)**.

Once launched, we'll start off by going into the **Display** tab (most likely, we'll already be in the **Display** tab) and verify that our settings in the **Plan**, **Model**, and **Section** views are as we would like them to appear.

Remembering that the name of this particular display style is **Contours 2' and 10' (Background)**, we're going to want to keep the **Major (10')** and **Minor (2')** contours toggled on, but may want to turn off the **Border** display, as shown in *Figure 4.12*:

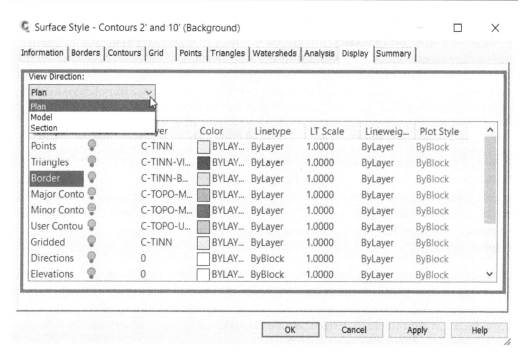

Figure 4.12 – Surface Style display settings

We also have the ability to change the **Layer**, **Color**, **Linetype**, **Linetype Scale**, **Lineweight**, and **Plot Style** options in the additional columns as needed. In the **Out of the Box** template that we imported containing the styles, all standard layers associated with these styles have been imported into our current file as well.

There may be some minor adjustments that you may want to implement with regard to the layer naming conventions, but it's important to note that the layers being imported, and associated with the styles themselves, are in alignment with layer naming conventions outlined in the **National CAD Standard** (**NCS**).

Although you may be required to make adjustments to the layer assignments and naming conventions, it's recommended that the additional columns (**Color**, **Linetype**, and so on) all remain as is to ensure that the controlling display is dictated by the layer assigned to those individual components and the associative properties.

Once we're comfortable with how our **Display** settings are being applied to our surface object in each view direction, we can make additional adjustments to the settings of the surface style by switching into any of the tabs listed along the top, as shown in *Figure 4.13*:

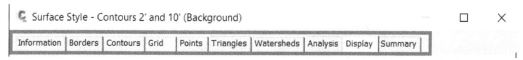

Figure 4.13 – Surface Style settings

The **Information** tab provides a brief overview of the name, description, and some additional details about the surface style. The **Borders**, **Contours**, **Grid**, **Points**, **Triangles**, **Watersheds**, and **Analysis** tabs allow for further defining some of the display settings associated with each component of the surface display. The **Summary** tab is essentially a quick access that displays all tabs and settings available throughout the various tabs, and also allows us to modify them in one consolidated location.

Now that we have a good idea of how we can update, modify, and refine our display settings associated with surfaces, we can apply similar concepts across all other Civil 3D-modeled objects within our Survey Model.

Going back to our Toolspace and selecting our **Prospector** tab, we can see that we also have **Alignments**, **Pipe Networks**, and **Pressure Networks** created in this file as well. If we were to select the modeled objects directly in our **Model** space, right-click, and select **Edit (object) Style**, we would be presented with similar dialog boxes that would allow us to further define and apply display settings associated with our objects.

With this section, we learned the importance of understanding our survey file so that we can next understand what we can do to customize it for our team's unique needs for creating our design further down the road. Survey files can vary greatly from group to group, and with Civil 3D, we have the ability to keep it in our company's design standards to make it easy to use when working, as well as producing final documentation down the road.

In the next section, we will begin actually working with the survey file and integrating it into our design documents, as our drawings will all begin to link and build upon each other. This is incredibly important, as taking the time to return to previous issues or missed information can be costly to a design team and can lead to faulty designs if not noticed soon enough.

Survey workflow overview

After going through and applying our various objects and settings within our Survey Model, we can now start to begin thinking about two major steps in our design workflow: beginning our design and syncing updates to our master project—or company—Civil 3D template.

To set up our Survey Model to be incorporated into our design models moving forward, we'll go ahead and follow the steps outlined in *Chapter 3, Sharing Data within Civil 3D* in the *Setting the working folder*, *Associating a project to the current drawing*, and, finally, *Creating data shortcuts* topics for all Civil 3D modeled objects in the Survey Model file.

From a Civil 3D template standpoint, we're going to want to make sure we continue to update our project-specific, or company-wide, Civil 3D template. That said, let's go ahead and perform a SAVEAS command (type this in the command line) to save our current file as a drawing template.

When the **Save Drawing As** dialog box appears, give it a descriptive name and change the **Files of type** option to **AutoCAD Drawing Template (*dwt)**, as shown in *Figure 4.14*:

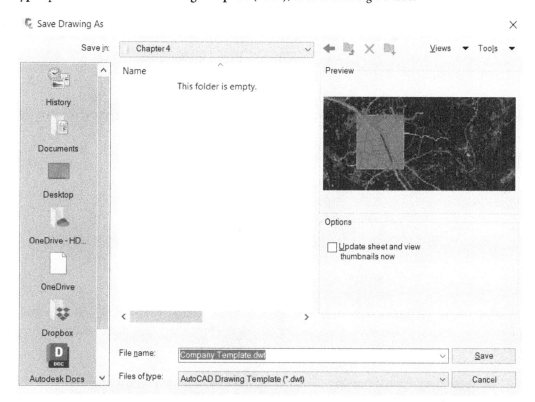

Figure 4.14 – Save Drawing As dialog box

After a descriptive name has been given and the **Save** button has been selected, we'll be presented with a **Template Options** dialog box where we can add some additional settings and a **Description** option for tracking purposes and to provide some additional clarification for the application/uses of the template, as shown in *Figure 4.15*:

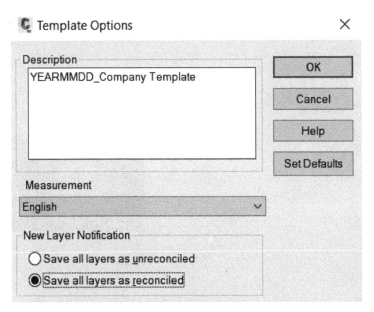

Figure 4.15 – Civil 3D Template Options dialog box

Once saved/created, you'll notice that the current file open is now your survey template, and no longer your Survey Model that we've been working on. We can tell this by looking back at the top of our Civil 3D session and noticing that the name indicated at the top now reflects the name of the file we just saved, with a .dwt extension, as shown in *Figure 4.16*:

Figure 4.16 – Current file open

The next step we'll take in our drawing template is to select everything in **Model** space (or press the *Alt + A* keys on your keyboard to select all), and then type ERASE at the command line (or simply hit the *Delete* key on your keyboard).

By doing this, we are essentially keeping our initial drawing template as clean as possible without containing any actual Civil 3D (or AutoCAD) objects in it. From a drawing template perspective, we really only want this file to contain the necessary display styles and settings that can be imported into our files moving forward.

For any updates we intend to make to the drawing templates from here on out, we'll take a slightly different approach by opening our drawing template, using the **Import Styles and Settings** workflow mentioned earlier in this chapter, and perform a SAVEAS to overwrite the file.

With this section, we have been able to utilize some of the incredibly powerful workflows of Civil 3D by referencing drawings from outside sources that utilize its styles and settings, but we can also frame them within our company's templates to display them as our team has agreed upon and to highlight features for ease of use in drawing creation, as well as plotting when the time comes to issue drawings.

This is not only highly favorable to our design process for working more efficiently, but it is also a large time saver versus previous ways that required redoing many of the elements manually and ensuring nothing was missed before moving on to the design.

Furthermore, you should have acquired a new skillset for customizing a survey file with your company's standards and templates without being at risk of modifying the existing conditions given to your group. This can be applied in later projects if needed for customers that require specific elements on specific layers or plotted in specific ways.

So, with that, multiple templates can be created and chosen as necessary as different projects from different clients arise, and you and your team can remain confident in your production level despite the varying projects being designed.

Analyzing your existing conditions

There are many ways we can begin inspecting and analyzing our existing conditions model. These methods can certainly be carried throughout our design lifecycle and applied to design models as well. Although we'll get into some more complex ways of analyzing our existing conditions model over the next few chapters, let's start out with some very simple methods that anyone just starting off can pick up.

Going back to our Survey Model_Start.dwg file, let's begin by taking a good look at our Surface model in a three-dimensional view to make sure that our existing surface at least appears to be accurate from a visual standpoint.

To perform this quick validation, we can simply select our existing Surface model in **Model** space by right-clicking and selecting the **Object Viewer…** option, as shown in *Figure 4.17*:

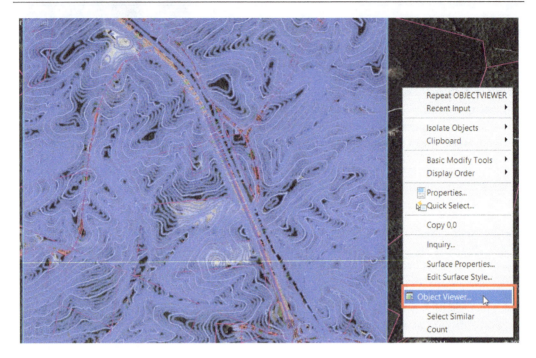

Figure 4.17 – Object Viewer selection

Once selected, an **Object Viewer** dialog box will appear where we can update **Visual Display Styles**, **Orientation**, and **Viewing Distances** options, as shown in *Figure 4.18*:

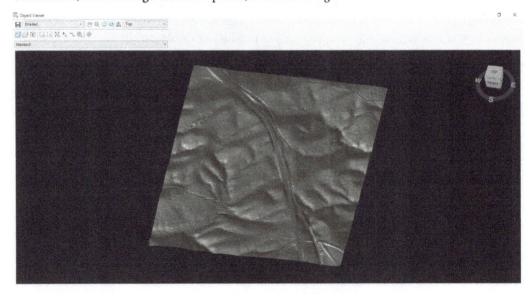

Figure 4.18 – Object Viewer with surface display

In the top left-hand corner of our **Object Viewer** dialog box, you'll notice a number of fields and tools available to us where we can fine-tune and modify our current display. On the top row, going from left to right, we have the following tools and options available to us (a detailed display of this row is shown in *Figure 4.19*):

- **Save Image**: Allows us to save/export the current view displayed in **Object Viewer** to an image file
- **Visual Styles**: Allows us to apply AutoCAD visual styles (preset and custom defined) to our current view displayed in **Object Viewer**
- **Pan Realtime**: Allows us to pan, or shift, our current view displayed in **Object Viewer**
- **Zoom Realtime**: Allows us to zoom in/out of our current view displayed in **Object Viewer**
- **Steering Wheel**: Allows us to quickly access many navigational tools to modify our current view displayed in **Object Viewer**
- **Constrained 3D Orbit**: Allows us to orbit three-dimensionally around our current view displayed in **Object Viewer**
- **Adjust Distance**: Allows us to adjust the perspective projection of our current view displayed in **Object Viewer** (note that this can only be adjusted if **Perspective Setting** is toggled on; refer to the second row of tools, also shown in *Figure 4.20*)
- **View Control**: Allows us to apply a preset view direction to our current view displayed in **Object Viewer**:

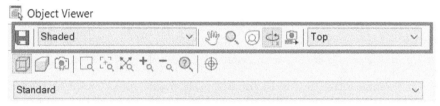

Figure 4.19 – Object Viewer with surface display

In the second row, going from left to right, we have the following tools and options available to us (a detailed display of this row is shown in *Figure 4.20*):

- **Parallel View**: Sets the current view displayed in **Object Viewer** to be parallel to our objects
- **Perspective View**: Sets the current view displayed in **Object Viewer** to be angled to our objects
- **Lens Length**: Sets a zoom factor to **Perspective View** (not available in **Parallel View**)
- **Zoom Window**: Allows us to zoom into our current view displayed in **Object Viewer** using the window option

- **Zoom Center**: Allows us to center our current view displayed in **Object Viewer**; this is done by clicking with our mouse on the intended center of our display

- **Zoom Extents**: Allows us to quickly zoom to the extents of all objects displayed in our current view in **Object Viewer**

- **Zoom In**: Allows us to zoom in to our current view displayed in **Object Viewer**

- **Zoom Out**: Allows us to zoom out in our current view displayed in **Object Viewer**

- **Zoom Factor**: Allows us to apply a zoom factor to our current view displayed in **Object Viewer**

- **Set View**: Allows us to apply our current view displayed in **Object Viewer** to our view being displayed in **Model** space:

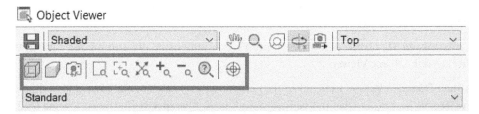

Figure 4.20 – Object Viewer with surface display

With that overview, we should feel comfortable moving forward with rotating and viewing our modeled geometry within **Object Viewer**. Using the tools discussed, feel free to navigate your model within the **Object Viewer** dialog box as you see fit. Performing a visual inspection can also be done directly in the **Model** space, but it is not recommended as the design files are large files and are very graphic intensive, which can cause your system to freeze or crash, thus risking a loss of work.

This is why it is recommended to perform a visual inspection with **Object Viewer** as it is much less graphicly intensive on your system. Some items that I typically look for when visually inspecting my Surface model would include the following:

- Verification of high/low points

- Breakline locations where curb lines, edge of pavements, retaining walls, shorelines, ditches/channels, and so on may be present on-site appear to be located accurately, and our Surface model reflects these more drastic changes in grade/slope/elevation

- Surface anomalies

- Holes or missing data

- All project data needed has been captured to complete the design properly

Once we've been able to visually inspect our modeled geometry, we can close out of **Object Viewer** and begin analyzing our pipe networks, pressure networks, and centerlines that reside in our Survey Model_Start.dwg file.

With that, let's go ahead and select our pipe networks, pressure networks, and Surface model all at once. To do so, instead of selecting these objects in our **Model** space, let's use the following steps to select these modeled objects:

1. Launch our Toolspace.
2. Select our **Prospector** tab.
3. Expand **Surfaces** and right-click on our existing surface.
4. Choose the **Select** option.
5. Expand **Pipe Networks**.
6. Expand **Networks**.
7. Right-click on our existing **Sanitary Sewer** Gravity Network.
8. Choose the **Select** option.
9. Right-click on our existing **Storm Drainage** Gravity Network.
10. Choose the **Select** option.
11. Expand **Pressure Networks** and right-click on **Existing Water**.
12. Choose the **Select** option.

Now that we have selected all three of these modeled objects from within our **Prospector** (**Surfaces**, **Pipe Networks**, and **Pressure Networks**), we can now right-click in **Model** space and select the **Object Viewer** option again, the same as we did earlier in this section while inspecting our Surface model alone.

As we pull up our **Object Viewer** now and navigate around our model, we can verify that our modeled geometry is located properly and there are no glaring conflicts that appear (visually) within our model at this time (refer to *Figure 4.21*):

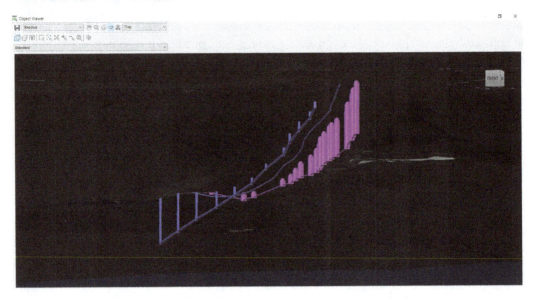

Figure 4.21 – Object Viewer with multiple objects in our display

When it comes to pipe networks and pressure networks, some items that I typically look for when visually inspecting in **Object Viewer** would include the following:

- Gravity pipe connections to structures

- Pressure pipe connections to fittings and appurtenances

- Presence of any clashing gravity and pressure pipes

- Quick inspection to verify that all gravity and pressure pipe networks are in fact under the Surface model as expected

- Gravity structures, and some pressure appurtenances, are at grade when viewing our Surface model as well

> **Note**
>
> It's important to note that we are not necessarily limited to viewing only one type of object at a time in our **Object Viewer**. We can certainly select everything in our drawing at any given time and view all 2D and 3D objects in our **Object Viewer**.

Now that we've had a chance to visually inspect and analyze our model a little bit, we should begin to feel a bit more comfortable with how our `Survey Model_Start.dwg` file has been constructed, along with the modeled geometry contained within.

Summary

At the present time, we have now established all the foundational components to successfully set up our project and are now prepared to begin our design. As we progress through the remaining chapters, we'll continue to revisit the steps of updating our project (or company) drawing template to make sure that we have an even sturdier foundation when we start our next project design.

Granted—in this example, we were in some ways afforded a bit of a lay-up in the sense that the Survey Model we started with had almost everything modeled and displayed as we would hope.

Many surveyors are very well versed in Civil 3D, and this is the type of Survey Model composition one can typically expect to receive. Unfortunately, that isn't always the case, and designers will occasionally be required to piece survey data together and construct the models within survey points.

With that, in our next chapter, we'll get a better understanding of how we can leverage points from a survey perspective, along with the **Line** and **Curve** tools that can be leveraged in parallel. Although these tools are not used very often in our day-to-day design workflows, this will give you an idea of some more of the powerful tools that are available to us with Autodesk Civil 3D.

This will also prepare us in the event that we will be required to compose our Survey Models from points alone.

5

Leveraging Points, Lines, and Curves

Now that we understand the basic tools and methods of Civil 3D, this chapter will dive into the application and intersection of these tools to better develop a confident infrastructure design. Civil 3D is a dynamic program with not only intelligent components and design processes but also integrated workflows that adjust when dependent parts of the design adjust or change, which results in less time spent recreating work or formatting when work is updated automatically.

In *Chapter 4, Configuring Survey Data with Civil 3D*, we reviewed many of the preparational tasks required to be performed in our Survey Model. In the previous chapter, we were afforded what I would consider a lay-up in the sense that survey data has already been converted into the 2D and 3D geometry making up our Survey Model.

Occasionally, we will come across instances where specific components will be resurveyed and then provided to the designer after the initial composition of the Survey Model in the form of an ASCII file, which is essentially a text-based file with survey points including **point numbers**, **northings**, **eastings**, **elevations**, and **descriptions**, also seen abbreviated as **PNEZD** or some similar variation.

With that, the following key topics will be covered in this chapter:

- Setting up a new file to import points from survey data
- Introduction to points
- Introduction to lines and curves

Throughout this chapter, we'll incorporate various point/line/curve commands to get an idea of how we can import survey and supplemental data relatively easily and convert it into geometry that can be incorporated into our overall Survey Model.

For this example, we're going to start with a brand-new file using the Civil 3D template we created in *Chapter 4, Configuring Survey Data with Civil 3D*, and use the ASCII file, both of which can be found in our `Practical Autodesk Civil 3D 2024\Chapter 5` subfolders entitled `Company Template File.dwt` and `Survey Model.asc`, respectively.

Technical requirements

The exercise files for this chapter are available at `https://packt.link/UoiPn`

Setting up a new file to import points from survey data

To kick things off, let's go ahead and create a new drawing using our new `Project/Company` drawing template file that we created, which can be found in our `Practical Autodesk Civil 3D 2024\Chapter 5` subfolder.

We'll open **Menu Browser** in the top-left corner of our Civil 3D session, and then select **New | Drawing**, as shown in *Figure 5.1*:

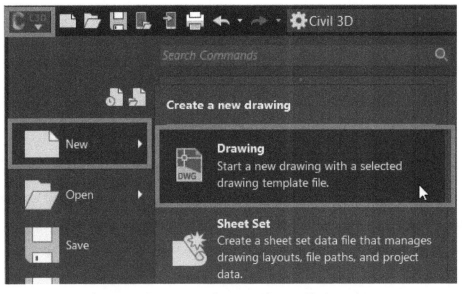

Figure 5.1 – Create a new drawing with a specified template

Once selected, we'll be asked to select the drawing template that we wish to create a drawing from, at which point, we'll navigate to our `Practical Autodesk Civil 3D 2024\Chapter 5` location, select the `Company Template File.dwt` file, and click **Open** (refer to *Figure 5.2*):

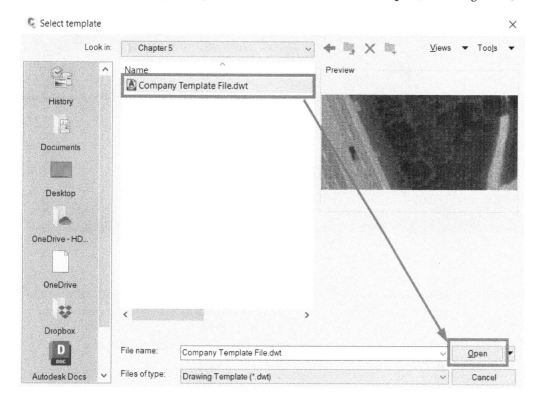

Figure 5.2 – Select drawing template

After selecting the file and clicking **Open**, a new drawing will be created that already incorporates the styles and settings that were applied to our Survey Model in *Chapter 4, Configuring Survey Data with Civil 3D,* prior to clicking **Save As** to create our drawing template. You'll notice right off the bat that our new drawing has been created to include **Aerial Imagery** as a background provided through the **Geolocation** ribbon.

This is due to the fact that when we saved our file as a drawing template, we saved the coordinate system projection as well. We can confirm this again by selecting the **Settings** tab in **Prospector** by right-clicking on our filename and selecting **Edit Drawing Settings...** to pull up the **Drawing Settings** dialog box that has the projection identified, as shown in *Figure 5.3*:

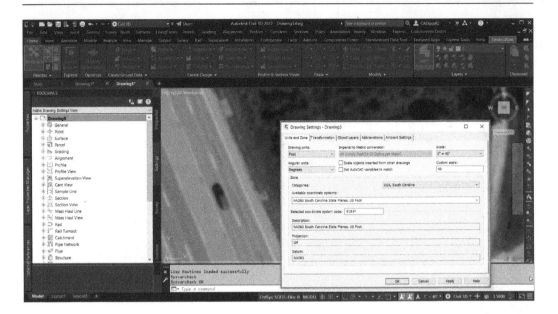

Figure 5.3 – The Drawing Settings dialog box

That said, if you tend to work on projects in multiple states, you can either update the setting so as not to include a coordinate system projection or begin to think about ways you can create multiple drawing templates for each state in which you will be performing the design. Either way, a few key things to think about when it comes to creating drawing templates include the following:

- We want to make sure that we have a great foundation to build off whenever we create new drawings and start new projects
- Make starting new drawings and projects as streamlined and automated as possible to ensure that the majority of time spent is on actual design and production
- Drive consistency through efficiency in all of your projects and designs

Once we're comfortable with the direction we're headed with the new drawing, and all the styles, settings, and our coordinate system assignment, let's go ahead and save our newly created drawing. To do so, we'll head back up to **Menu Browser**, hover over **Save As**, and select **Drawing**, as shown in *Figure 5.4*:

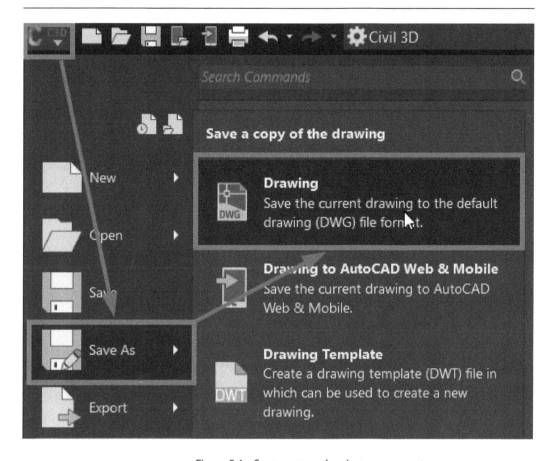

Figure 5.4 – Save our new drawing

Once selected, we'll be presented with a **Save Drawing As** dialog box, at which point we'll navigate to our Practical Autodesk Civil 3D 2024\Chapter 5 location, call the Survey Model.dwg file, and click on the **Save** button, as shown in *Figure 5.5*:

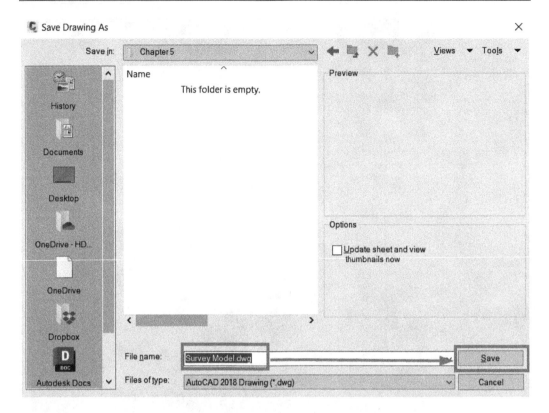

Figure 5.5 – Save our new drawing

Now that we have our file set up and saved, we can next move on to understanding how we can incorporate survey data via ASCII point files into our Survey Model for future use.

Introduction to points

Points in Civil 3D can have many different applications, purposes, and nomenclature. They are often viewed as one of the most basic foundational components that can be used for the creation of models and design progression and for painting the story of our overall design.

For the purposes of this chapter, we'll focus on the importing process of points generated from our surveyed data, and identify some beneficial applications where these imported points can be processed to generate geometry and models to be displayed in our `Survey Model.dwg` file.

Starting with the importing process, we'll be able to access the `import Points` command by going up to the ribbon, selecting the **Insert** tab, and then selecting the **Points from File** tool, as shown in *Figure 5.6*:

Figure 5.6 – Import points from file

Once selected, we'll be presented with the **Import Points** dialog box where we'll go ahead and import the Survey Model ASCII file into our `Survey Model.dwg` file using the following steps, as numbered in *Figure 5.7*:

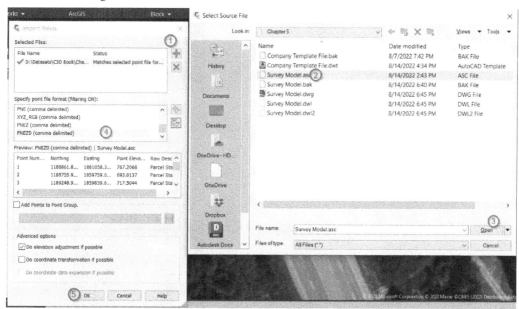

Figure 5.7 – The Import Points dialog box

1. Click on the **Add Files** icon.

2. Select the `Survey Model.asc` file located in `Practical Autodesk Civil 3D 2024\Chapter 5`.

3. Click the **Open** button.

4. Select **PNEZD** as the point file format (**PNEZD** represents **Point Number**, **Northing**, **Easting**, **Elevation**, **Z Value/Elevation**, and **Description**).

5. Click the **OK** button.

> **Note**
>
> Importing may take a little while in this instance as there are 36,294 points to import from the `Survey Model.asc` file. Civil 3D can accept a variety of point files such as `.txt`, `.csv`, and more. It can also accept them in a few different formats such as comma-delineated or space-delineated, depending on how the point file was created in the field.

After all our points have been imported, we not only see them displayed in our model space but also listed in our panorama if we select **Points** in **TOOLSPACE | Prospector**, as shown in *Figure 5.8*:

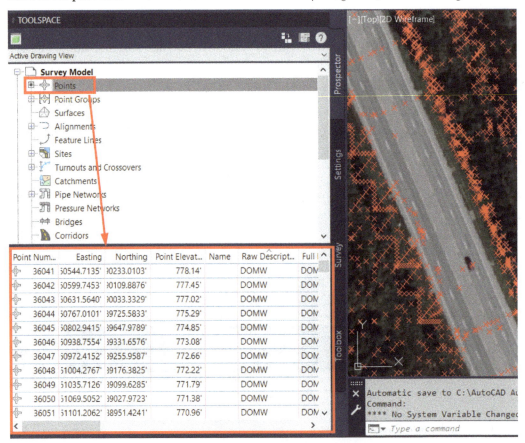

Figure 5.8 – Points displayed in the panorama

If you were to navigate through the panorama, you would notice that each of the points had a raw and full description associated with them as well. By using these descriptions, we can start to organize and add filterable selections by parsing our points out and generating individual point groups based on the descriptions.

It is important to note that the points imported here are referred to as **coordinate geometry (COGO) points**. These points are more directly editable than other points in Civil 3D such as survey points, which are linked directly to a survey database and are more difficult to adjust if need be.

With that, let's go ahead and click the + icon next to **Point Groups** to expand it out a bit. We notice that there is currently a point group already available labeled **_All Points**. It is important to note that the **_All Points** point group is included by default and cannot be deleted from our drawings. On the flip side, any new point groups that are created can be deleted as needed.

In any event, let's go ahead and add a few organizational levels to our points in our current file. To do so, we'll right-click on **Point Groups** in **Prospector** and select **New...**, as shown in *Figure 5.9*:

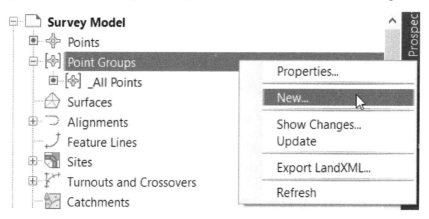

Figure 5.9 – Create new point group

Once selected, we'll be presented with the **Point Group Properties** dialog box where we can begin to further define and organize the points that we inserted into our file. In the **Information** tab, let's go ahead and give our first point group a proper name, description, and default point and label styles as listed next, and shown in *Figure 5.10*:

- **Name**: Ground
- **Description**: Existing Ground Survey Shots
- **Default Point Style**: <none>
- **Default Point Label Style**: <none>

It is important to note that the <none> point style and point label style will assure us that these points are not displayed in our view. However, we can still use these to generate our 2D and 3D geometry moving forward.

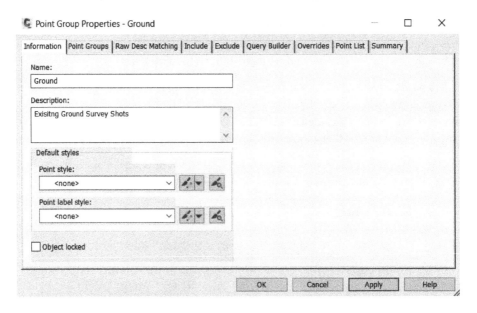

Figure 5.10 – Point Group Properties information tab

Next, we'll switch over to the **Include** tab, check the box next to **With raw descriptions matching**, and type *GRND* into the field next to this option, as shown in *Figure 5.11*. The * symbol is used as a wildcard to match and include any points that contain GRND in the raw description of the points currently residing in the file:

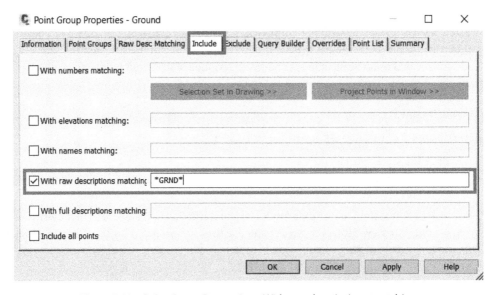

Figure 5.11 – Point Group Properties – With raw descriptions matching

Now, if we switch over to the **Point List** tab and click on the **Apply** button in the lower right, we'll see a fully comprehensive list of all the points in our file that include `GRND` in the raw description, as shown in *Figure 5.12*:

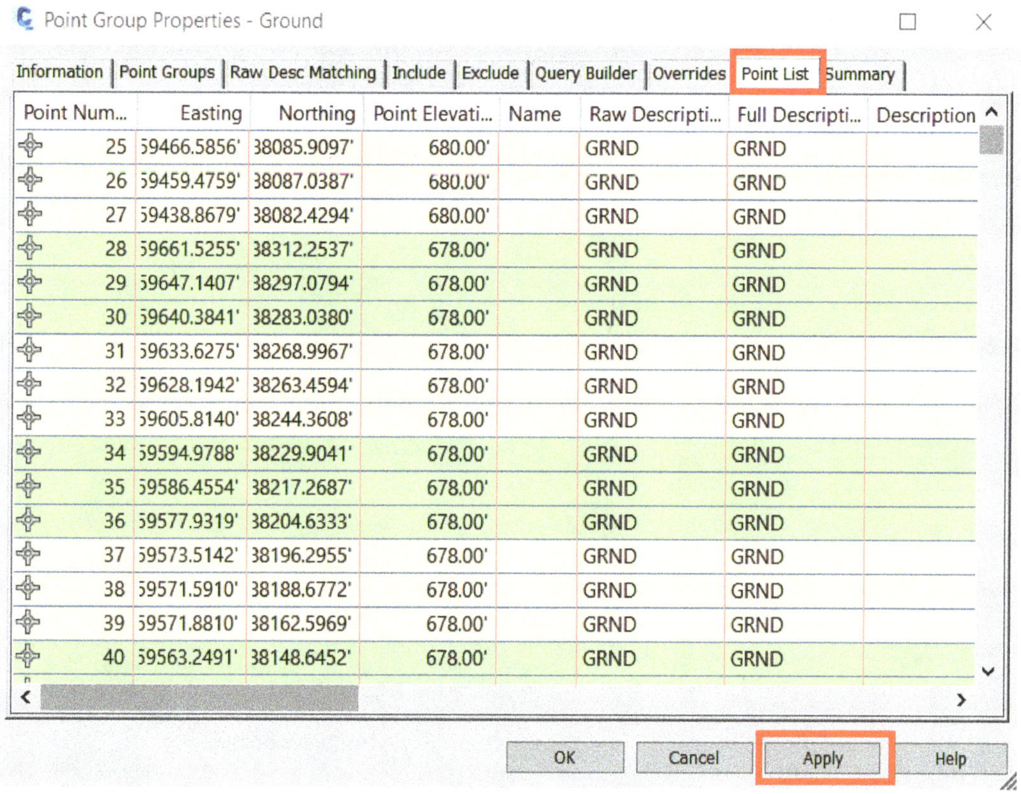
Point Group Properties - Ground
Information
Point Groups
Raw Desc Matching
Include
Exclude
Query Builder
Overrides
Point List
Summary
Point Num... Easting Northing Point Elevati... Name Raw Descripti... Full Descripti... Description
25 59466.5856' 38085.9097' 680.00' GRND GRND
26 59459.4759' 38087.0387' 680.00' GRND GRND
27 59438.8679' 38082.4294' 680.00' GRND GRND
28 59661.5255' 38312.2537' 678.00' GRND GRND
29 59647.1407' 38297.0794' 678.00' GRND GRND
30 59640.3841' 38283.0380' 678.00' GRND GRND
31 59633.6275' 38268.9967' 678.00' GRND GRND
32 59628.1942' 38263.4594' 678.00' GRND GRND
33 59605.8140' 38244.3608' 678.00' GRND GRND
34 59594.9788' 38229.9041' 678.00' GRND GRND
35 59586.4554' 38217.2687' 678.00' GRND GRND
36 59577.9319' 38204.6333' 678.00' GRND GRND
37 59573.5142' 38196.2955' 678.00' GRND GRND
38 59571.5910' 38188.6772' 678.00' GRND GRND
39 59571.8810' 38162.5969' 678.00' GRND GRND
40 59563.2491' 38148.6452' 678.00' GRND GRND
OK
Cancel
Apply
Help

Figure 5.12 – Point Group Properties – Point List

Next, we'll move on to our **Sanitary Sewer Manhole** points. As we did during the first point group creation, we'll right-click on **Point Groups,** and select **New** again. In the **Information** tab of the **Point Group Properties** dialog box, we'll go ahead and fill out the **Name** (`Sanitary Sewer Manholes`) and **Description** (`Existing Sanitary Sewer Manholes`) fields.

We'll select **Sanitary Sewer Manhole** as **Point style** and then apply **<none>** as **Point label style**, as shown in *Figure 5.13*:

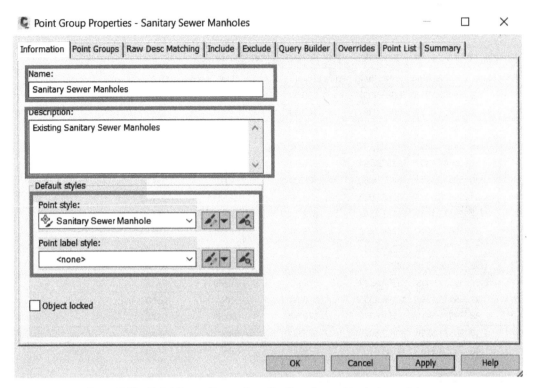

Figure 5.13 – Point Group Properties – Sanitary Sewer Manholes point group

Moving over to the **Include** tab, we'll want to check the box next to **With raw description matching** and type `*SSWR*` into the field next to this option, and then click on the **Apply** and **OK** buttons.

For **Storm Drainage**, we'll take a slightly different approach by going into the **Point Groups** section in **Prospector**, right-clicking on our **Sanitary Sewer Manholes** point group that we just created, and selecting **Copy...**, as shown in *Figure 5.14*:

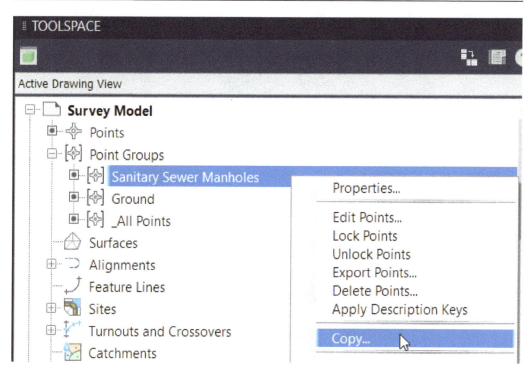

Figure 5.14 – Copy point group

You'll notice that a new point group called **Copy of Sanitary Sewer Manholes** is now displayed in our list of point groups. To update and change this to apply to **Storm Drainage Structures**, we'll go ahead and right-click on the copied point group and select **Properties...**, as shown in *Figure 5.15*:

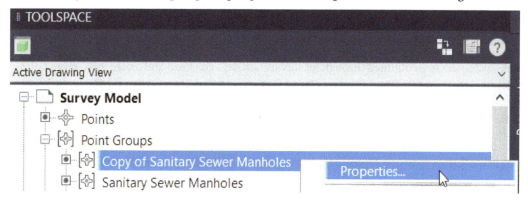

Figure 5.15 – Edit point group properties

In the **Information** tab, we'll go ahead and fill out the **Name** and **Description** fields and apply default styles, as shown in *Figure 5.16*:

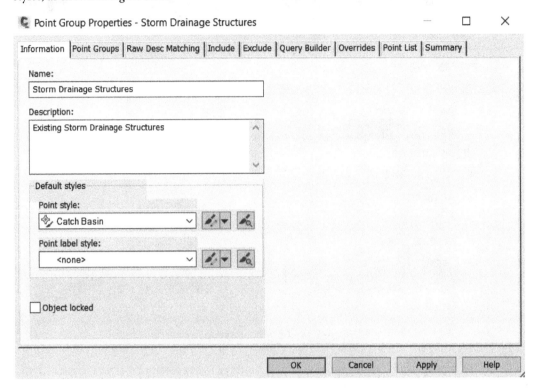

Figure 5.16 – Point Group Properties – Storm Drainage Structures

Then, we'll switch over to the **Include** tab and change **With raw description matching** from *SSWR* to *STRM* in the field next to this option, and then click on the **Apply** and **OK** buttons.

Finally, using the steps outlined throughout this section, let's go ahead and create three more point groups to cover **Domestic Water Pipeline**, **Parcel Staking**, and **York Hwy CL**. For the **Domestic Water Pipeline** and the **York Hwy CL** point groups, we'll apply the **<none>** point style and point label style. For the **Parcel Staking** point group, we'll want to apply the **Iron Pin Point** style.

In the end, we should only see **Sanitary Sewer Manholes** and **Storm Drainage Structures** in the model space in our file, as shown in *Figure 5.17*. It is important to note that we can control the appearance of points with point groups as well as point styles.

This itself can be strategically used further in design and labeling, but be sure to understand there is still a control hierarchy where point groups are the parent display controlling how all points contained within are set, and then individual point styles themselves will apply to those not within point groups.

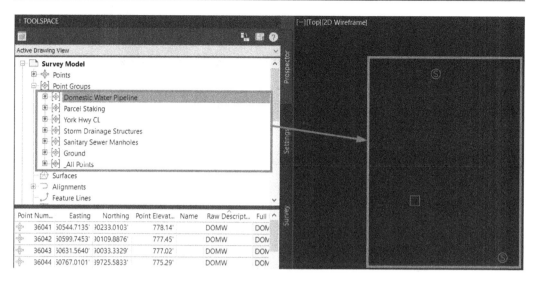

Figure 5.17 – Final display after point groups are created

Although it may not seem like much based on what we can currently see in our model space, we have actually done a tremendous service by setting ourselves up to leverage some additional tools available to us within Civil 3D to streamline the generation of a full survey display. These tools and workflows will be discussed throughout the remainder of this chapter.

Introduction to lines and curves

Now that we have our points imported into our file, we can start to think about ways to leverage the line and curve commands to generate some of our basic survey geometry. In the previous section, we created several point groups in an effort to organize the survey data that we received and imported into our Survey Model.

In our first exercise, we'll go ahead and convert the existing road centerline points into real linework. With that, let's go ahead and identify the point number range within the **York Hwy CL** point group (these points are meant to depict the centerline of York Highway).

To identify the point range, we'll go back into our **TOOLSPACE | Prospector** tab, select the point group labeled **York Hwy CL**, and take note of the point numbers in the panorama listed, as shown in *Figure 5.18*:

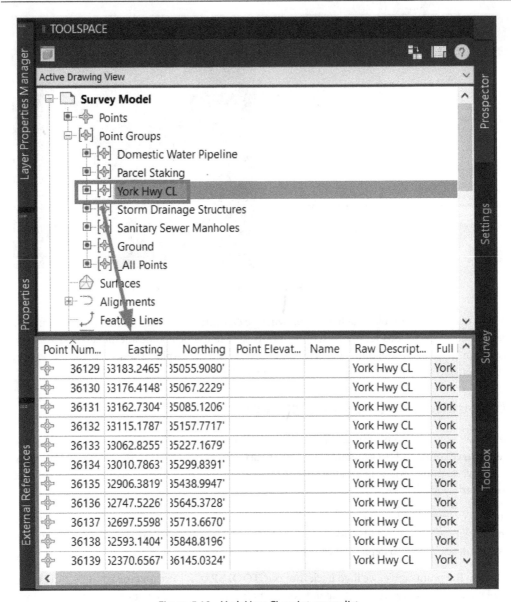

Figure 5.18 – York Hwy CL point group list

We'll notice that the point numbers start with **#36,083** and end with **#36,294**.

Let's now move over to the **Home** ribbon, zoom to the **Draw** panel, and select the down arrow next to the **Line** icon to display all the available options we have to generate lines, as shown in *Figure 5.19*:

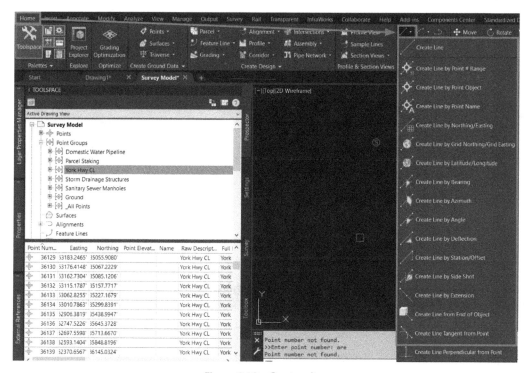

Figure 5.19 – Create a line

As you can see, this list is quite extensive and provides a multitude of ways for us to generate lines in our current file. For additional details about each method, we can simply hover our mouse over each method, at which point a detailed overview will be displayed explaining how each can be applied.

For the purposes of this exercise, let's focus on those methods that allow us to incorporate points into our line generation:

- **Create Line by Point # Range**: We can generate lines by inputting the range of points that we'd like to connect between

- **Create Line by Point Object**: We can generate lines by selecting the points that we'd like to connect between

- **Create Line by Point Name**: We can generate lines by inputting the point name associated with the points we'd like to connect between

In our case, since we do not have a point name applied to our points, and we have already set our **York Hwy CL** point group display to **<none>**, we are left to use the **Create Line by Point # Range** command.

Let's go ahead and select that command in our **Home** | **Draw** | **Lines** pull-down menu. We'll notice at the very bottom of our session that we will be asked to enter the point number. Instead of typing in 104 point numbers manually, we can simply type in the point range that we took note of earlier, for instance, we're using a point range of 35217-35321 (as shown in *Figure 5.20*) and then hit the *Enter* key on the keyboard:

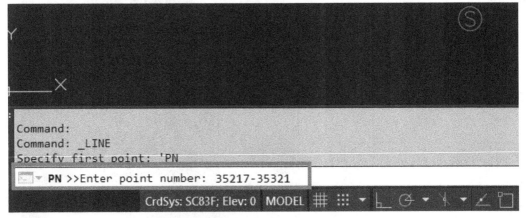

Figure 5.20 – Enter point numbers at the command line

Once the command has been processed, we'll use the following steps to convert our new linework into a polyline (refer to *Figure 5.21*):

1. Zoom out to the point where we have all the new linework within our view in the model space.

2. Go to the **Modify** ribbon.

3. Go to the **Modify** panel.

4. Click the down arrow in the **Modify** panel to expand the tools.

5. Select the **Join** command to convert individual lines into one continuous 3D polyline.

6. Select all lines.

7. Hit the *Enter* key on the keyboard:

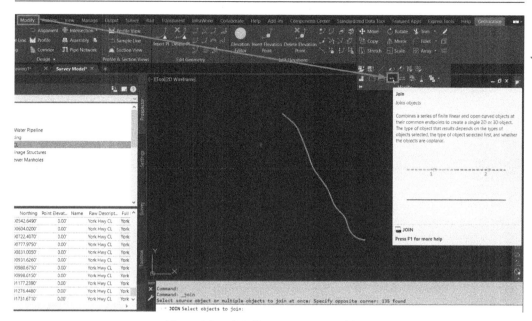

Figure 5.21 – Join lines to create a polyline

Next, we'll use the following steps to place our newly created polyline on the correct survey layer (refer to *Figure 5.22*):

1. Select the polyline we just created.

2. Go to the **Home** ribbon.

3. Go to the **Layers** panel.

4. Select the down arrow next to the current layer name.

5. Scroll down and select the **V-ROAD-CNTR** layer:

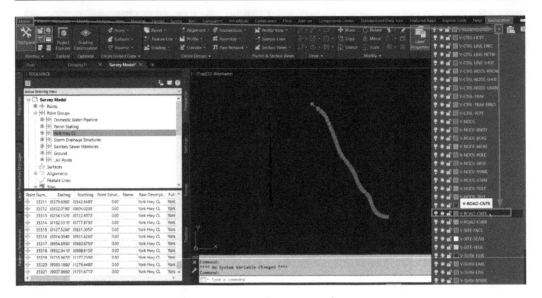

Figure 5.22 – Place polyline on corresponding layer

We have now officially converted our first set of points into geometry within our Survey Model. By converting our data into actual geometry, we are building the existing conditions, or built environment, from which we will be able to generate our design.

Let's continue to proceed with converting our utility point groups using the same workflow and then place them on their corresponding layers as follows:

- **Storm Drainage Structures**:

    - Using point range 35,163-35,174

    - Use the **Join** command to create one continuous 3D polyline

    - Place the resulting linework on the **V-STRM-MAIN** layer

- **Sanitary Sewer Manholes**:

    - Using point range 35,175-35,193

    - Use the **Join** command to create one continuous 3D polyline

    - Place the resulting linework on the **V-SSWR-MAIN** layer

- **Domestic Water Pipeline**:

    - Using point range 35,194-35,215

    - Use the **Join** command to create one continuous 3D polyline

    - Place the resulting linework on the **V-WATR-MAIN** layer

Moving on to our **Parcel Staking** point group, we'll use the **Create Line by Point Object** command (refer to *Figure 5.23*). Using this command, we will be prompted to select a start point and an end point.

Once selected, a line will be automatically generated to connect the start and end points. Before starting this command, we'll want to set our current layer to **V-PROP-LINE** so that all the new linework we create during this process will be created on the correct layer:

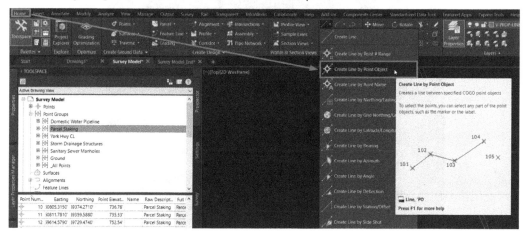

Figure 5.23 – Create Line by Point Object

Starting with the southern parcel, we'll select points to generate our parcel boundary, as shown in *Figure 5.24*:

Figure 5.24 – Existing parcel (southern)

Next, we'll perform the same task on our northern parcel, with the finished result shown in *Figure 5.25*:

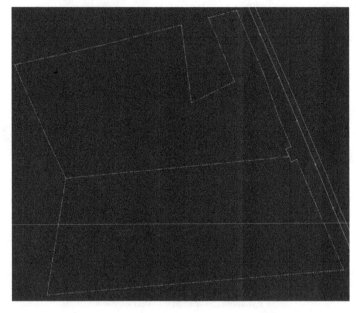

Figure 5.25 – Existing parcels

With the exception of the **Ground** point group, we have now converted all remaining point groups into real geometry within our Survey Model depicting various components of the existing built environment.

As I'm sure you've noticed, we have placed a heavy emphasis on the line commands as we are able to use these commands in conjunction with our surveyed points to generate our linework. The curve commands can be accessed within the **Home** ribbon in the **Draw** panel, similar to the line commands, as shown in *Figure 5.26*:

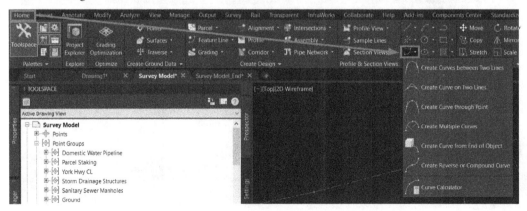

Figure 5.26 – Curve commands

As displayed in *Figure 5.26*, we do not have the ability to generate curves from our point groups automatically, but we do have the ability to create curves:

- **Create Curve between Two Lines**: Allows us to add a fillet to two selected lines

- **Create Curve on Two Lines**: Adds a fillet but leaves the selected lines untouched

- **Create Curve through Point**: Creates a fillet by selecting lines and defining the location for the curve to run through

- **Create Multiple Curves**: Creates multiple compound curves between selected lines with varying lengths and radii

- **Create Curve from End of Object**: Creates an arc as a continuation of the selected line

- **Create Reverse or Compound Curve**: Creates either a reverse or compound curve as a continuation of the selected arc

Although we do not currently have a workflow identified where these commands can be utilized, we will be revisiting these tools and applying them in workflows identified in *Chapter 7, Alignments - The Second Foundational Component to Designs within Civil 3D*, and *Chapter 10, Roadway Modeling Tool Belt for Everyday Use*.

Summary

We are now well on our way to fully understanding how we can set up a Survey Model and establish a solid foundation from which we can build our design. In the event that the surveyor is required to go out into the field to survey additional data, we have a good understanding and should feel a little more at ease if we are to add additional survey data and convert linework to be displayed in our Survey Model.

In our next few chapters, we'll continue to build up our Survey Model to incorporate 3D elements for which we can data-reference as needed and develop our design models. In the next chapter, in particular, we'll use the **Ground** point group that we left untouched to understand how we can build, manage, and analyze our Surface Models.

Surfaces - The First Foundational Component to Designs within Civil 3D

Within any Civil 3D design, surfaces, alignments, and profiles are often considered the three major foundational components required to build a true Civil infrastructure design model. In *Chapter 4*, *Configuring Survey Data with Civil 3D*, we were afforded the benefits of having our survey data already being converted to 2D and 3D geometry, making up our survey model.

In *Chapter 5*, *Leveraging Points, Lines, and Curves*, we reviewed many of the preparational tasks that we need to perform to process surveyed data into 2D geometry to be displayed in our survey model.

In this chapter, we'll begin by focusing on surfaces as the root foundation 3D component that we will use to reference and build additional 3D components from. That said, key topics that will be covered in this chapter include the following:

- Generating a surface model
- Understanding surface styles
- Surface manipulation and management

Leveraging the survey point data and point groups we created in *Chapter 5*, *Leveraging Points, Lines, and Curves*, we'll pick up where we left off by opening the `Survey Model.dwg` file located in the `Practical Autodesk Civil 3D 2024\Chapter 6` subfolder.

Technical requirements

The exercise files for this chapter are available at `https://packt.link/UoiPn`

Generating a surface model

If you recall, in *Chapter 5*, *Leveraging Points, Lines, and Curves*, we created point groups for each of our survey points and processed them into linework, with the exception of our **Ground Point Group**. In this section, we'll use **Ground Point Group** to generate a 3D surface model that we will use as a reference for building our design.

To kick things off, we'll start by going into **TOOLSPACE**, selecting the **Prospector** tab, and locating the **Surfaces** category. If we right-click on **Surfaces**, we will be presented with the following options, also displayed in *Figure 6.1*:

- **Create Surface…**: Provides the ability to create a new surface model (TIN, Grid, TIN Volume, or Grid Volume)

- **Create Surface From DEM…**: Provides the ability to create a new surface model (via USGS **Digital Elevation Model (DEM)**), GEOTIFF, ESRI ASCII Grid, or ESRI Binary Grid)

- **Create Surface from TIN…**: Provides the ability to create a new surface model from a TIN file

- **Show Preview**: When Show Preview is enabled, or checked, a preview of the individual modeled object will display at the bottom of our Toolspace dialog box

- **Rebuild Out of Date Items**: Rebuilds surface models if a component applied to the surface definition has been updated

- **Create Folder**: Provides the ability to manage surfaces at a level deeper by adding folder creation capabilities

- **Export to DEM**: Exports all surface models to any of the following formats: DEM and GEOTIFF

- **Export to LandXML**: Exports all surface models to a LandXML file

- **Refresh**: Refresh

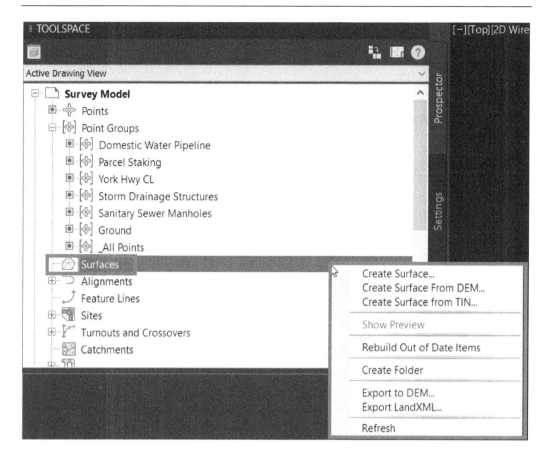

Figure 6.1 – Right-click options for Surfaces

To ensure that we are keeping our surface models organized, we'll go ahead and select the **Create Folder** option. In the **Create Folder** dialog box, set **New folder name** to Existing Conditions, as shown in *Figure 6.2*:

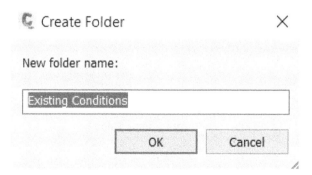

Figure 6.2 – Create Folder dialog box

Going back to the **Surfaces** category, right-click again on **Surfaces** but select the **Create Surface** option this time. When the **Create Surface** dialog box appears, we'll make the following selections (also shown in *Figure 6.3*):

- **Type**: **TIN surface** (the type of surface model we wish to create).

- **Surface Layer**: **C-TOPO** (the layer on which our surface model as a whole will be placed).

- **Name**: **SRF - Existing Grade – FromSurveyPoints** (naming conventions are important to define at the beginning to drive consistency and make things clear as to what each component represents. In this case, **SRF** represents **Surface, Existing Grade – FromSurveyPoints** and indicates that this particular component represents the overall existing surface model and has been generated from survey point data.

- **Description**: **Created from Ground Point Group** (it's always good practice to add a description and clarifying statements as to how the surface model was constructed).

- **Style**: **Contours 2' and 10' (Background)** (represents the style that will be applied to our surface model that will appear in our plan views).

- **Render Material**: **ByLayer** (represents the style that will be applied to our surface model that will appear in 3D views).

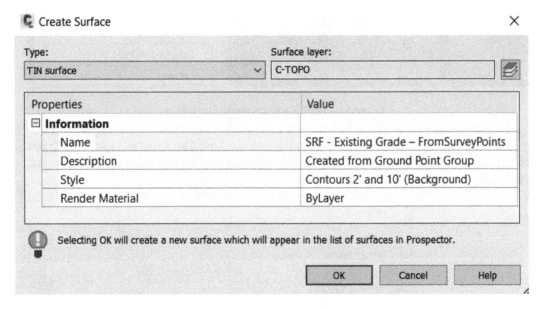

Figure 6.3 – Create Surface dialog box

Once all the fields have been filled out accordingly, click the **OK** button. Going back to the **TOOLSPACE | Prospector** tab, let's go ahead and select our newly created surface and drag and drop it into the `Existing Conditions` folder, as shown in *Figure 6.4*:

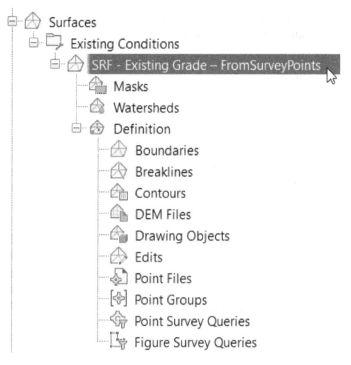

Figure 6.4 – Create Surface dialog box

Click on the + icon next to the **SRF - Existing Grade – FromSurveyPoints** surface model along with the + icon next to **Definition**. With these expanded, we can now begin adding components to our existing surface model. As we add components, these will update in our model space, so keep an eye out for these updates occurring.

As it relates to components, we are able to add to the surface definition here. We have the following options available to us:

- **Boundaries**: Allows us to apply **Outer**, **Show and Hide**, and **Data Clip** boundaries to our surface models

- **Breaklines**: Allows us to apply **Standard**, **Proximity**, **Wall**, **From File**, and **Non-Destructive Breaklines** to our surface models

- **Contours**: Allows us to apply existing polylines representing manually placed contours to our surface models

- **DEM Files**: Allows us to apply external **USGS DEM**, **GEOTIFF**, **ESRI ASCII Grid**, or **ESRI Binary Grid** files to our surface models

- **Drawing Objects**: Allows us to apply **AutoCAD Points**, **Lines**, **Blocks**, **Text**, **3D Faces**, and **Polyfaces** to our surface model

- **Edits**: Allows us to perform various edits to our surface model (that is, modifying **TIN lines**, **Adding/Deleting Points**, **Minimizing Flat Areas**, **Smoothing Surfaces**, and **Simplifying Surfaces**)

- **Point Files**: Allows us to apply an external **Point File** (in `.txt`, `.prn`, `.csv`, `.auf`, `.nez`, `.pnt` format) to our surface model

- **Point Groups**: Allows us to apply **Point Group** to our surface model

- **Point Survey Queries**: Allows us to apply and generate a dynamic link to survey points (typically used in surveying workflows)

- **Figure Survey Queries**: Allows us to apply and generate a dynamic link to survey figures (typically used in surveying workflows)

For the purposes of our **SRF - Existing Grade – FromSurveyPoints** surface model, we want to apply **Point Groups** to our surface definition. That said, let's right-click on the **Point Groups** option and select **Add**, as shown in *Figure 6.5*:

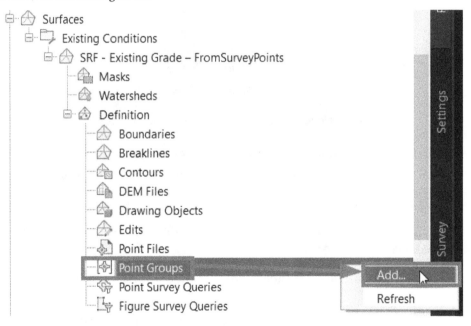

Figure 6.5 – Add Point Groups to surface definition

We'll then be presented with a **Point Groups** dialog box where we can then select the point groups created in our current survey model file from which we can build our **SRF - Existing Grade – FromSurveyPoints** surface model.

In our case, we'll obviously want to select the **Ground** point group, click **Apply**, and then **OK**, as shown in *Figure 6.6*:

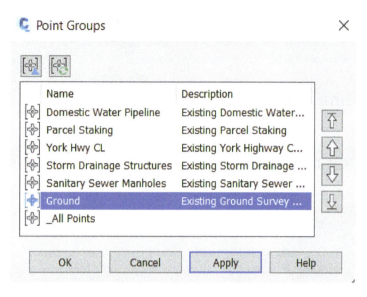

Figure 6.6 – Point Groups dialog box

After clicking **OK** in the **Point Groups** dialog box, we'll notice that the **SRF - Existing Grade – FromSurveyPoints** surface model has now appeared in our model space and is displayed using the **2' and 10' (Background)** display style we applied at the initial creation, as shown in *Figure 6.7*:

Figure 6.7 – SRF - Existing Grade – FromSurveyPoints surface model

We have now officially created our very first surface model within Civil 3D and have a great foundation from which to build our design! We can now begin exploring different ways we can view, analyze, manipulate, and manage our surface model moving forward.

Understanding surface styles

Importing our survey data and creating our surface models for sheet preparation is one thing (which we accomplished in the previous section). Understanding what we're looking at by applying different surface styles is a completely different ballgame.

In this section, we'll familiarize ourselves with the various surface styles available to us and learn how to display our surfaces under different circumstances.

For this section, let's leave the `Survey Model.dwg` file and open the `Survey Model_ DisplayStyles_Start.dwg` file located in the `Practical Autodesk Civil 3D 2024\ Chapter 6` subfolder. Once it's open, you will notice that it should look very similar to where we left off in the `Survey Model.dwg` file.

> **Note**
> Since we will be running various tests on the new file we have opened, it's best practice to leave the original as is to avoid introducing any type of file corruption. The common practice for setting up these types of testing environments is to simply perform a *Save As* (type this on the command line) and save it as the same name with `Working` at the end. Adding this suffix to the drawing name will notify users that this file is for testing purposes and not meant to be a final drawing.

In the current `Survey Model_DisplayStyles_Start.dwg` file, let's go over to **TOOLSPACE**, select the **Settings** tab, and expand the **Surface** category by clicking the + icon next to **Surface**.

Notice that we have the following list of subcategories associated with surfaces (also displayed in *Figure 6.8*):

- **Surface Styles**: Here, we can view the comprehensive list of available display styles associated with surface models.

- **Label Styles**: Here, we can view the comprehensive list of available label styles that can be applied to our surface models for annotation purposes; these can also display specified symbology.

- **Table Styles**: Here, we can view the comprehensive list of available table styles that can be displayed for sheeting purposes as they relate to our surface models.

- **Commands**: Here, we can view the comprehensive list of available commands that can simply be typed into the command line to perform specific tasks associated with surfaces.

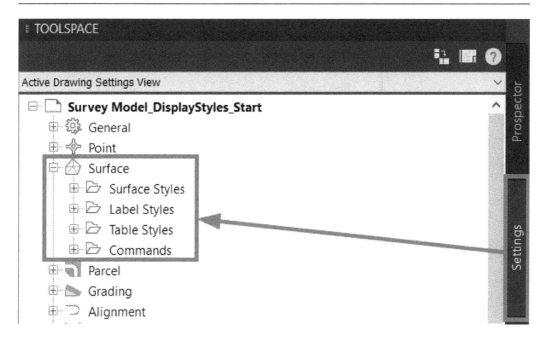

Figure 6.8 – TOOLSPACE | Settings tab – surface styles expanded

Starting with surface styles, if we click on the + icon next to **Surface Styles**, we will then see the comprehensive list of options that reside in our current drawing.

> **Note**
>
> These are also contained within the Company Template File.dwt drawing template that we created as a starting point for new files. All styles listed from here on will be included in any new file, provided you are using the Company Template File.dwt drawing template as the base.

Once expanded, we'll see the following list of surface styles that can be applied to our surface models, within our current file, for display purposes (also refer to *Figure 6.9*):

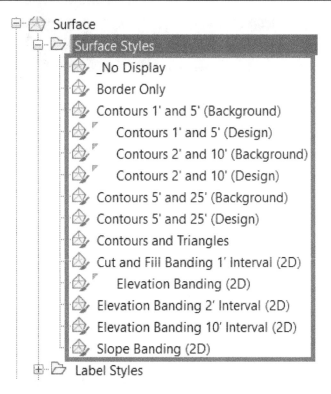

Figure 6.9 – Surface Styles

- **_No Display**: Applying this display to our surface model will turn all components off in model space.

- **Border Only**: Applying this display to our surface model will only turn on the border(s) within model space.

- **Contours 1' and 5' (Background)**: Applying this display to our surface model will only turn on minor and major contours at 1' intervals in model space. Contours displayed will be assigned to layers that have grey assignments, so they are screened when plotted.

- **Contours 1' and 5' (Design)**: Applying this display to our surface model will only turn on minor and major contours at 1' intervals in model space. Contours displayed will be assigned to layers that have brighter color assignments so they are more prominent when plotted.

- **Contours 2' and 10' (Background)**: Applying this display to our surface model will only turn on minor and major contours at 2' intervals in the model space. Contours displayed will be assigned to layers with gray assignments, so they are screened when plotted.

- **Contours 2' and 10' (Design)**: Applying this display to our surface model will only turn on minor and major contours at 2' intervals in the model space. Contours displayed will be assigned to layers that have brighter color assignments so they are more prominent when plotted.

- **Contours 5' and 25' (Background)**: Applying this display to our surface model will only turn on minor and major contours at 5' intervals in the model space. Contours displayed will be assigned to layers with gray assignments, so they are screened when plotted.

- **Contours 5' and 25' (Design)**: Applying this display to our surface model will only turn on minor and major contours at 5' intervals in the model space. Contours displayed will be assigned to layers that have brighter color assignments so they are more prominent when plotted.

- **Contours and Triangles**: Applying this display to our surface model will turn on the triangles and border(s), along with the minor and major contours at 2' intervals in model space. Elements displayed will be assigned to layers with gray assignments so they are screened when plotted.

- **Cut and Fill Banding 1' Interval (2D)**: Applying this display to our surface model will only turn on elevations, where the color scheme displayed in model space is initially identified by the default settings applied in **Surface Style | Analysis | Elevations | Range color scheme** (refer to *Figure 6.10*).

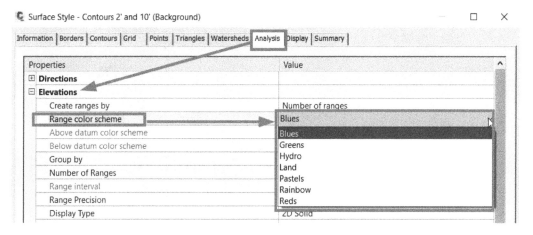

Figure 6.10 – Surface Style properties | Analysis tab | Elevations default settings

- **Elevation Banding (2D)**: Applying this display to our surface model will only turn on elevations, where the color scheme displayed in the model space is initially identified by the default settings applied in **Surface Style | Analysis | Elevations | Range color scheme** (refer to *Figure 6.10*).

- **Elevation Banding 2' Interval (2D)**: Applying this display to our surface model will only turn on elevations, where the color scheme displayed in the model space is initially identified by the default settings applied in **Surface Style | Analysis | Elevations | Range color scheme** (refer to *Figure 6.10*).

- **Elevation Banding 10' Interval (2D):** Applying this display to our surface model will only turn on elevations, where the color scheme displayed in model space is initially identified by the default settings applied in **Surface Style | Analysis | Elevations | Range color scheme** (refer to *Figure 6.10*).

- **Slope Banding (2D):** Applying this display to our surface model will only turn on elevations, where the color scheme displayed in the model space is initially identified by the default settings applied in **Surface Style | Analysis | Elevations | Range color scheme** (refer to *Figure 6.10*).

Next, we'll visit surface label styles by clicking on the + icon next to **Label Styles** – at which point, we'll see a comprehensive list of surface label styles that reside in our current drawing.

Once expanded, we'll see the following list of **Label Styles** that can be applied to our surface models within our current file for annotation display purposes (for a more detailed breakdown of **Label Styles** in our current file, refer to *Figure 6.11*):

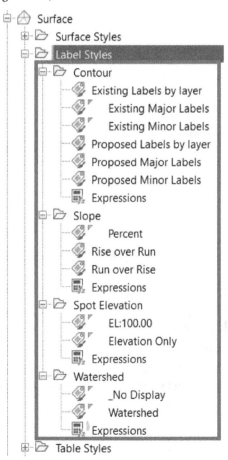

Figure 6.11 – Surface Label Styles (detailed version)

- **Contour**: Contour labels are displayed using the **Add Surface Labels** option within the **Annotate** ribbon (refer to *Figure 6.12* for the workflow and location), provided a **Surface Display Style** showing contours (**Minor** and/or **Major**) are displayed as well. Please note that surface contour labels will automatically be associated with the contour interval defined in the **Surface Display Style** settings.

- **Slope**: Slope labels will be displayed in various formats as we identify locations where we would like to place them using the **Slope** labels option within the **Annotate** ribbon (refer to *Figure 6.12* for workflow and location).

- **Spot Elevation**: Spot labels will be displayed in various formats as we identify locations where we would like to place them using the **Spot Elevation** option within the **Annotate** ribbon (refer to *Figure 6.12* for workflow and location).

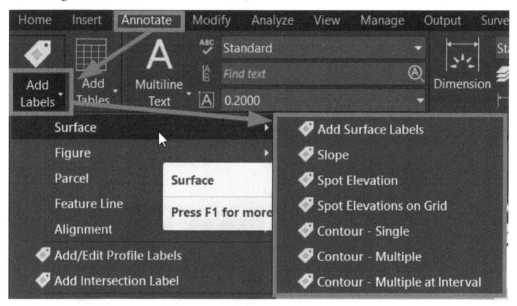

Figure 6.12 – Workflow for adding surface labels

- **Watersheds**: Watershed labels will be displayed as defined in the **Watershed Label Properties**, provided a **Surface Display Style** showing the watersheds is displayed as well, where the label styles displayed in model space are initially identified by the default settings applied in **Surface Style | Watersheds | Surface | Surface Watershed Label Style** (refer to *Figure 6.13*).

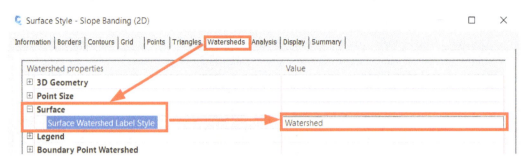

Figure 6.13 – Workflow for adding surface labels

Next, we'll look at the surface table styles by clicking on the + icon next to **Table Styles** – at which point, we'll then see a comprehensive list of surface table styles that reside in our current drawing as well (refer to *Figure 6.11*).

> **Note**
>
> It's important to note that surface table styles will reference the user-defined settings applied in the **Analysis** tab in the **Surface Properties** dialog box (refer to *Figure 6.14* for the location and example settings).

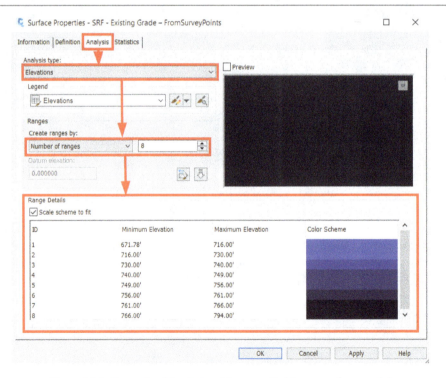

Figure 6.14 – Example of Analysis settings applied to our surface model

Once expanded, we'll see the following list of **Table Styles** (also refer to *Figure 6.15* for a detailed breakdown of the surface table styles in our current file) that can be added to our drawings for sheeting purposes.

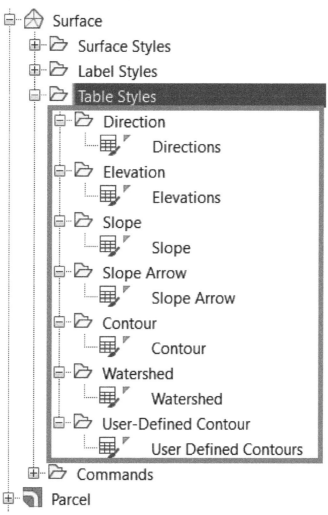

Figure 6.15 – Surface Table Styles (detailed version)

- **Direction**: This table style will be dynamically linked to the settings defined in *Figure 6.14* and display direction ranges, colors, and area values accordingly

- **Elevation**: This table style will be dynamically linked to the settings defined in *Figure 6.14* and display elevation ranges, colors, area, and volume values accordingly

- **Slope**: This table style will be dynamically linked to the settings defined in *Figure 6.14* and display slope ranges, colors, and face area values accordingly

- **Slope Arrow**: This table style will be dynamically linked to the settings defined in *Figure 6.14* and display slope arrow ranges, colors, and face area values accordingly
- **Contour**: This table style will be dynamically linked to the settings defined in *Figure 6.14* and display contour ranges, colors, areas, and volume values accordingly
- **Watershed**: This table style will be dynamically linked to the settings defined in *Figure 6.14* and display number, type, description, color, hatching, and drainage area values accordingly
- **User-Defined Contour**: This table style will be dynamically linked to the settings defined in *Figure 6.14* and display user-defined contour ranges, colors, areas, and volume values accordingly

With the overview of all applicable styles that can be applied to our surfaces out of the way, let's get a better understanding of what some of these actually mean and represent, how we can apply different styles under different circumstances and scenarios, and enhance our overall understanding of an existing conditions surface model, and any other surface model for that matter.

Surface manipulation and management

In the majority of cases, we're going to want to display our surface model with contours displayed, where typical intervals applied would be either 1' (minor) and 5' (major) or 2' (minor) and 10' (major), depending on the complexity and the size of the site where the design will be developed from. As we work through our designs, we may want to switch our surface model display as different design elements are introduced.

For example, say we want to perform a slope analysis of the existing surface model to quickly identify any potential risks or concerns related to slope stabilization on our current site. Although there are other factors that play into these types of analysis (such as soil types, water content, and vegetation), we can get a pretty good idea of what the current conditions are by performing a simple slope analysis on our existing surface model.

To do so, let's jump back into the Survey Model_DisplayStyles_Start.dwg file, select our surface model displayed in model space, right-click and select the **Surface Properties** option, as shown in *Figure 6.16*:

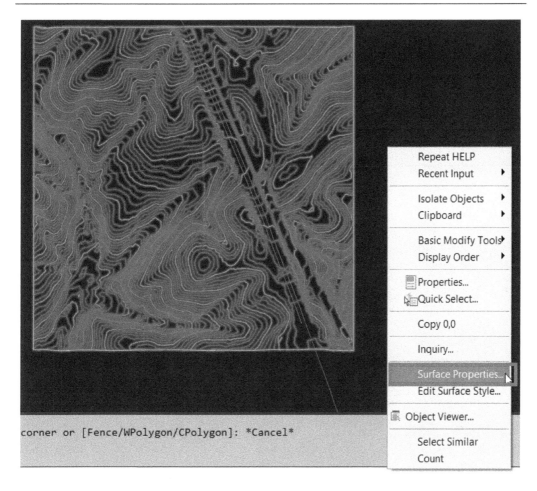

Figure 6.16 – Access Surface Properties in the model space

Once the **Surface Properties** dialog box appears, switch over to the **Analysis** tab and apply the following settings to begin our slope analysis of our existing surface model:

- **Analysis type: Slopes**

- **Legend: Slope**

- **Ranges | Number: 8**

- Click the down arrow icon next to our **Ranges | Number**

- Adjust **Minimum Slope** and **Maximum Slope** in **Range Details** as appropriate

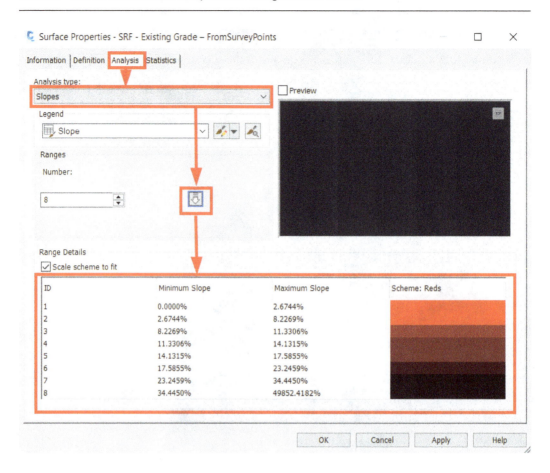

Figure 6.17 – Surface Properties dialog box | Analysis tab – applying slope analysis

Once we're comfortable with our settings, we'll then switch back over to the **Information** tab in the **Surface Properties** dialog box and apply the **Slope Banding (2D)** surface style (refer to *Figure 6.18*), and click on the **OK** button in the bottom right corner of the dialog box.

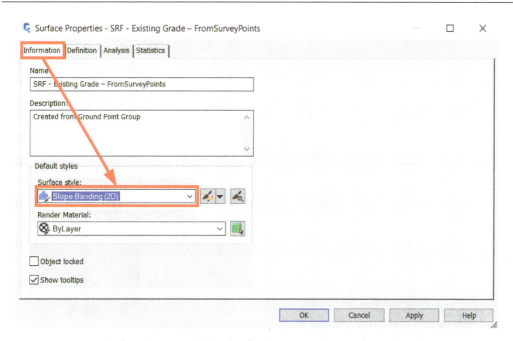

Figure 6.18 – Surface Properties dialog box | Information tab – applying slope display style

After clicking on the **OK** button to exit the **Surface Properties** dialog box, we will now see that the display of our surface has changed from displaying as **2' and 10' Contours** to **Slope Banding (2D)** across the entire surface model, as shown in *Figure 6.19*:

Figure 6.19 – SRF - Existing Grade – FromSurveyPoints slope analysis

If you are not seeing the survey linework we generated in previous exercises after changing the **SRF - Existing Grade – FromSurveyPoints** surface model display to **Slope Banding (2D)**, then you will need to change the display order of the elements within the file. To do so, simply select the **SRF - Existing Grade –FromSurveyPoints** surface model in model space, right-click, hover your mouse over the **Display Order** option, and finally select the **Send to Back** option, as shown in *Figure 6.20*:

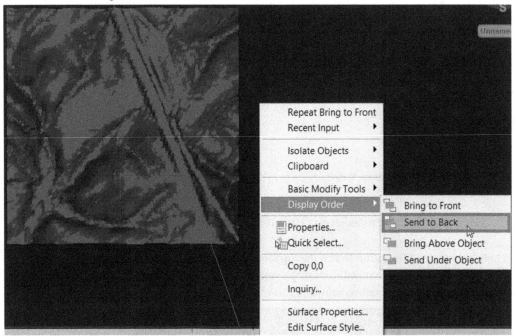

Figure 6.20 – Update Display Order of SRF - Existing Grade – FromSurveyPoints

As identified in *Figure 6.21*, there are several areas on our site where steeper slopes are present. It's good to note these areas and ensure that as we progress through our design, we are able to stabilize these areas to withstand long-term weather events and support the construction of our design with minimal impact on the surrounding areas.

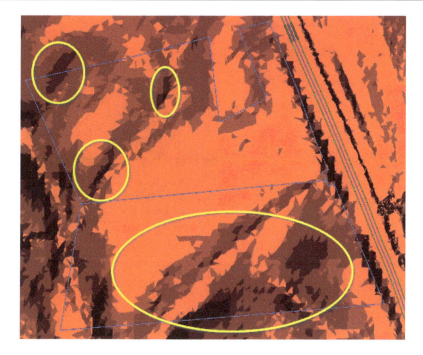

Figure 6.21 – Steep slopes within SRF - Existing Grade – FromSurveyPoints

Next, analysis that comes in handy as we prepare for our design is identifying elevation ranges, high and low spots on our site, and ridges and valleys to understand how and where our existing built environment is draining during storm events.

To perform this type of analysis, select the **SRF - Existing Grade – FromSurveyPoints** surface model again in the model space, right-click and select the **Surface Properties** option, just as we did when we started to perform our slope analysis previously.

Once the **Surface Properties** dialog box has appeared, we'll switch back over to the **Analysis** tab and apply the following settings to begin our elevation analysis of our existing surface model (refer to *Figure 6.22* for an image of settings):

- **Analysis type: Elevations**
- **Legend**: **Elevations**
- **Create ranges by | Number of ranges**: **8**
- Click the down arrow icon below the **Number of ranges** field
- Adjust **Minimum Elevation** and **Maximum Elevation** in **Range Details**, as appropriate

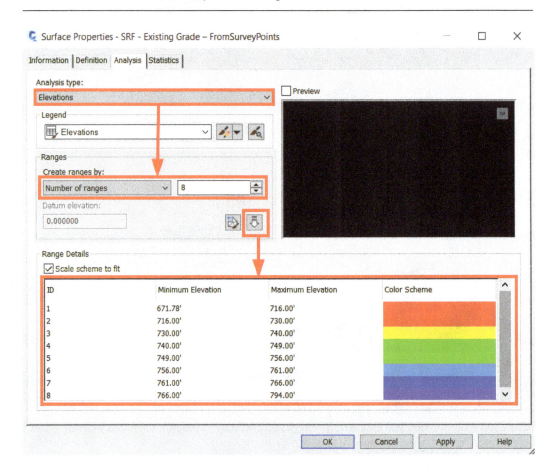

Figure 6.22 – Surface Properties dialog box | Analysis tab – elevation analysis

Once we're comfortable with our settings, we'll then switch back over to the **Information** tab in the **Surface Properties** dialog box and apply the **Elevation Banding (2D)** surface style, as shown in *Figure 6.23*, and click on the **OK** button in the bottom right corner of the dialog box.

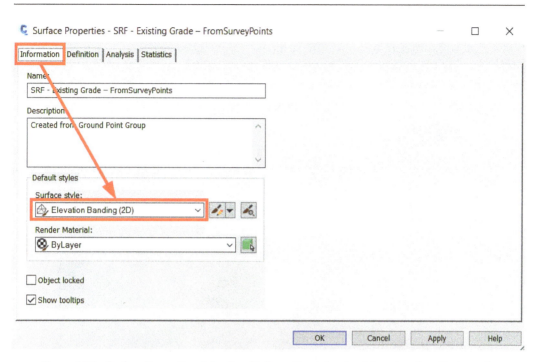

Figure 6.23 – Surface Properties dialog box | Information tab – Applying elevation display style

After clicking on the **OK** button to exit the **Surface Properties** dialog box, we can now see that the display of our surface has changed from displaying as **Slope Banding (2D)** to **Elevation Banding (2D)** across the entire surface model, as shown in *Figure 6.24*:

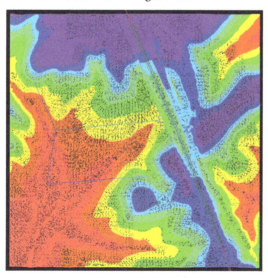

Figure 6.24 – Elevations and slope within SRF - Existing Grade – FromSurveyPoints

As identified in *Figure 6.25*, our existing site appears to be gradually sloping down toward the back of our site, away from York Highway. We can tell this by quickly identifying the high and low spots in our surface model when the **Elevation Banding (2D)** display style is applied. Furthermore, we can also identify how and where the existing site is draining water during storm events.

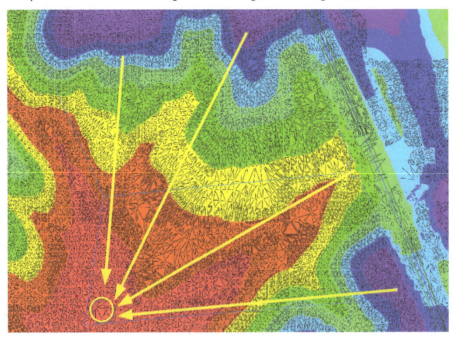

Figure 6.25 – SRF - Existing Grade – FromSurveyPoints drainage slopes

Taking this information into account, we'll need to maintain similar drainage patterns as we progress through our design. By maintaining similar drainage patterns, we are minimizing disruption and any impact on surrounding properties and areas.

Furthermore, as we work through our design and think about the actual construction of it, we'll need to apply similar tactics during the required construction phasing and staging and when designing and applying erosion control measures to our site.

Now that we have performed some initial investigation and interrogation of our existing surface model, giving us a decent idea of what we're up against from a design standpoint, we can switch the **SRF - Existing Grade – FromSurveyPoints** surface model back to the **Contours 2' and 10' (Background)** display style.

We can do this by selecting our existing surface model again in the model space, right-clicking, and selecting the **Surface Properties** option one more time. Once the **Surface Properties** dialog box appears, we'll go ahead and switch our display style from the **Elevation Banding (2D)** display style to the **Contours 2' and 10' (Background)** display style and click **OK**, as shown in *Figure 6.26*:

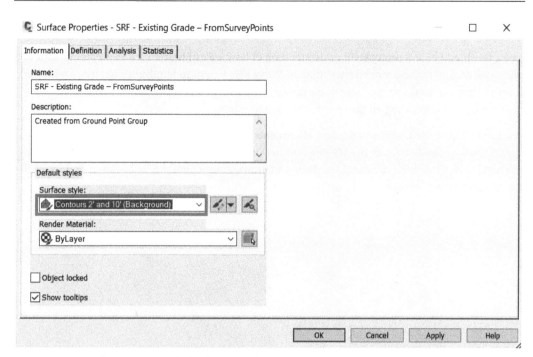

Figure 6.26 – Surface Properties dialog box – applying the Contours 2' and 10' (Background) display style

Following the steps outlined in *Chapter 3*, *Sharing Data within Civil 3D*, the last step we'll take is to create a new data shortcut project, associate it with our current drawing, and then add our **SRF - Existing Grade – FromSurveyPoints** surface model so that we can later data reference for design purposes into our working models.

As shown in *Figure 6.27*, you'll notice that when we create the data shortcut for our existing surface model, the subfolder entitled `Existing Conditions` that we created earlier in this chapter is automatically created as well.

As projects become more complicated and more team members are involved in project designs, this added level of organization in how we manage our modeled elements can be critical to keeping designs on track and under budget.

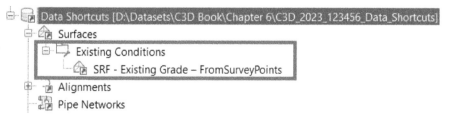

Figure 6.27 – Create the Data Shortcuts project and add the SRF -
Existing Grade – FromSurveyPoints surface model

Summary

By now, we should have a pretty solid understanding of how we can best utilize survey point data to create, analyze, and manage our surface models. We've also begun to familiarize ourselves with the various styles associated with surfaces in particular and can begin thinking about the different ways we can apply each depending on the circumstances and intent.

In the next chapter, we'll begin to explore alignments, including an overview of the various ways to create, manage, and apply all the styles associated with alignment geometry. We'll also begin to understand how alignments are considered a foundational component that is common and integral to the majority of our designs.

7
Alignments - The Second Foundational Component to Designs within Civil 3D

As mentioned, within any Civil 3D design, Surfaces, Alignments, and Profiles are often considered the three major foundation components required to build a true civil infrastructure design model. In *Chapter 6*, *Surfaces - The First Foundational Component to Designs within Civil 3D*, we took a deep dive into surfaces. We learned about several workflows to create, modify, interrogate, and analyze the existing Surface Model that we had generated from survey point data.

In this chapter, we'll take a similar deep dive into the world of alignments. Not only will we explore alignments in our existing Survey but we'll also begin to explore how, in many design scenarios, our design models can—and may need to—start with an alignment.

In our design scenario, we will be taking two existing large lots of land and splitting them into smaller parcels of land to essentially generate a residential subdivision. In addition to distributing our parcels, we'll need to design a road, right-of-way, and easements/setbacks for our houses to be placed on, along with proper drainage and various utilities to serve each property.

With that, in this chapter, we'll focus on the following key aspects as they relate to alignments in Civil 3D:

- Alignment creation
- Understanding alignment styles
- Alignment manipulation and management

As discovered with surfaces in *Chapter 6*, *Surfaces - The First Foundational Component to Designs within Civil 3D*, there are many levels of each foundation 3D geometry to peel to fully understand how integral a part they will play in our overall design. With that, let's jump right in to continue the momentum we already have going for us.

Technical requirements

The exercise files for this chapter are available at `https://packt.link/UoiPn`

Alignment creation

In this section, we'll begin exploring the multiple ways to create an alignment, all while getting a full understanding of the alignment creation tools available to us within the Civil 3D environment. To kick things off, we'll go ahead and open up our `Survey Model.dwg` file located in our `Practical Autodesk Civil 3D 2024\Chapter 7` subfolder.

For design purposes, we'll need to convert our linework depicting the centerline of York Hwy CL we created back in *Chapter 5*, *Leveraging Points, Lines, and Curves*. We'll need to convert to an alignment so that we have something to tie our design alignment into when we start laying out our subdivision itself. By tying two alignments to each other, we're able to dynamically design our intersections that will represent the entrance(s) of our site.

Moving over into our `Survey Model.dwg` file, let's go ahead and isolate our linework depicting York Hwy CL. To do so, we'll select our linework, right-click, and select **Isolate Objects | Isolate Selected Objects** (refer to *Figure 7.1*):

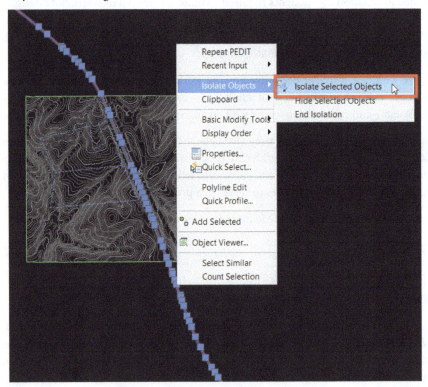

Figure 7.1 – Isolate Selected Objects workflow

Once our linework has been isolated (as shown in *Figure 7.2*), we'll notice that the linework depicting York Hwy CL is actually a 3D polyline. If you'll recall, when we originally created this geometry, we used our survey points to generate this linework.

That said, our points, in addition to having *X* and *Y* values (Northings and Eastings), also have *Z* values (elevations) associated with them as well. Since elevation ranges can vary when creating linework from points, this will, by default, convert the linework to a 3D polyline when we use the JOIN command (a detailed workflow was outlined in *Chapter 5*, *Leveraging Points, Lines, and Curves*:

Figure 7.2 – Isolated York Hwy CL

When converting linework to an alignment within the Civil 3D environment, we are limited to the types of objects that we are able to convert to an alignment. As noted in *Figure 7.3*, object types that we are able to convert to an alignment include lines, curves, and polylines:

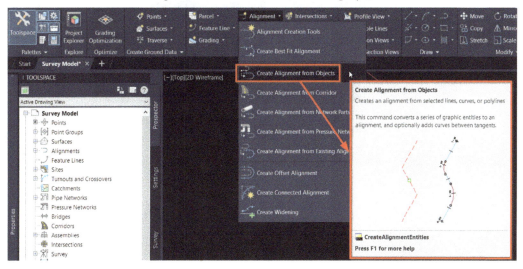

Figure 7.3 – Objects that can be converted to an alignment

> **Note**
>
> It's important to note that whenever Civil 3D references polylines, the implied object is actually either a 3D polyline or a 2D polyline. When 3D polylines are able to be used to convert to Civil 3D modeled elements, it will specifically call out 3D polylines.

Understanding that with the York Hwy CL linework currently shown in our file as a 3D polyline and that we are unable to convert 3D polylines to an alignment, we'll need to find an alternative solution for converting our linework to an alignment. Luckily for us, Civil 3D does come equipped with a tool that will allow us to convert 3D polylines to 2D polylines, and vice versa.

This tool can be found by selecting the **Modify** ribbon, expanding the design panel, and selecting the **Convert 3D to 2D Polylines** tool, as shown in *Figure 7.4*:

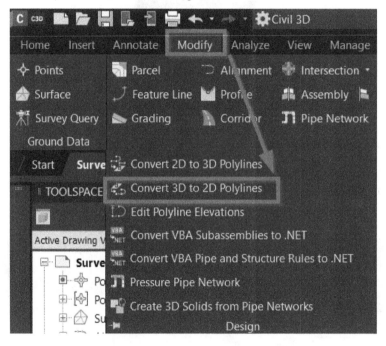

Figure 7.4 – Convert 3D to 2D Polylines

Once this tool has been initiated, we'll follow the steps detailed at the command line to continue with our conversion process. In our case, we'll be prompted to select 3D polylines to convert, at which point, we can go ahead and simply select our linework depicting York Hwy CL and hit the *Enter* key to accept.

The conversion process has now been applied to our linework. If we were to select the linework again, right-click, and select **Properties**, we'd be able to quickly see that our elected object is no longer being listed as a 3D polyline but as a polyline instead, as shown in *Figure 7.5*:

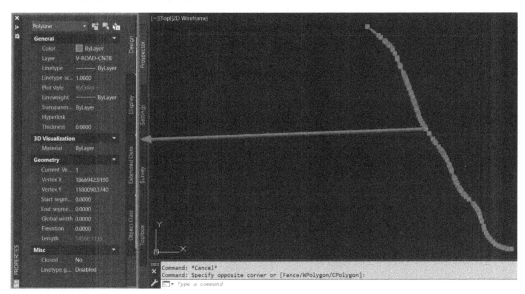

Figure 7.5 – Converted linework from a 3D polyline to a polyline

With our linework converted to a polyline, we can now convert our polyline to an alignment. To do this, we'll switch back to our **Home** ribbon, select the down arrow next to **Alignment** in our design panel, and then select the **Create Alignment from Objects** option, as shown in *Figure 7.6*:

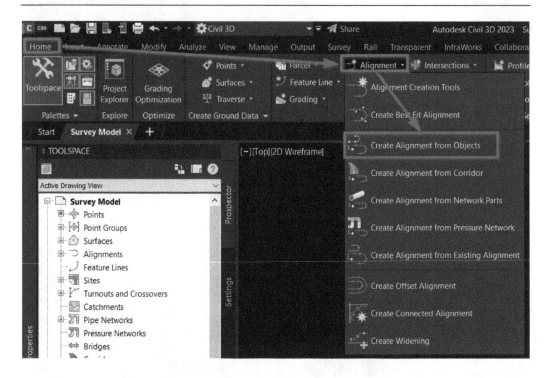

Figure 7.6 – Create Alignment from Objects

Once **Create Alignment from Objects** has been selected, we'll be prompted at the command line with the following steps:

1. **Select lines/arcs or polylines to create alignment**: We can go ahead and select our polyline depicting York Hwy CL.

2. **Press Enter to accept alignment direction or [Reverse]:** This option sets our direction and the stationing for our alignment. If we type *R* and hit the *Enter* key, this will reverse the direction of our alignment and stationing.

After getting through these steps, we'll then be presented with a **Create Alignment from Objects** dialog box, where we will fill out and make the following assignments (also shown in detail in *Figure 7.7*):

- **Name**: ALG - Existing York Hwy - FromSurveyPoints
- **Type**: **Centerline**
- **Alignment style**: **Existing**
- **Alignment label set**: **Major and Minor only**

- **Conversion options**:

    - **Add curves between tangents**: Uncheck this box to convert linework as is

    - **Erase existing entities**: Check the box to erase the polyline and replace it with an **Alignment** object in Civil 3D

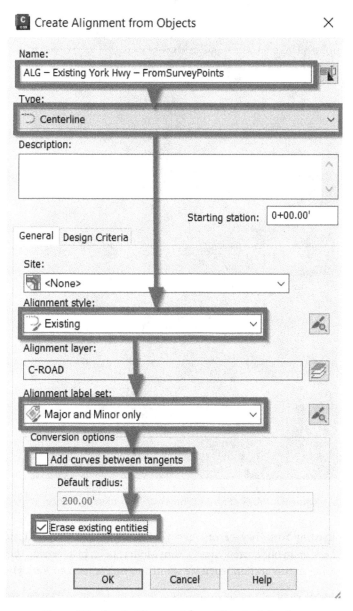

Figure 7.7 – Create Alignment from Objects dialog box

After making the appropriate selections and clicking the **OK** button, let's zoom in on our new alignment shown in Model Space. As we zoom in, we can see that the original linework has automatically been removed, with an intelligent Civil 3D alignment created in its place, containing check marks at **50'** intervals and station labels at **100'** intervals (refer to *Figure 7.8* for a zoomed-in view):

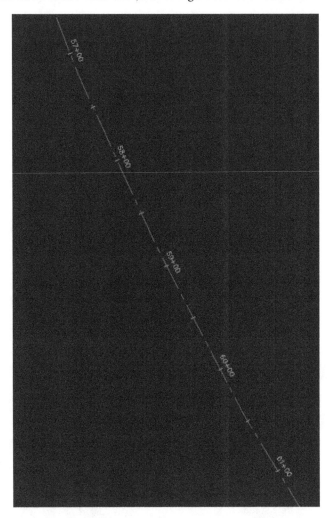

Figure 7.8 – ALG – Existing York Hwy – FromSurveyPoints

Now that our **ALG – Existing York Hwy – FromSurveyPoints** alignment has been created, let's go ahead and turn everything back on in our current view within Model Space, by reversing the object isolation that we performed earlier in this section when we wanted to isolate just our York Hwy CL linework.

To do this, we can simply right-click our mouse anywhere in our Model Space and select **Isolate Objects | End Object Isolation**, as shown in *Figure 7.9*:

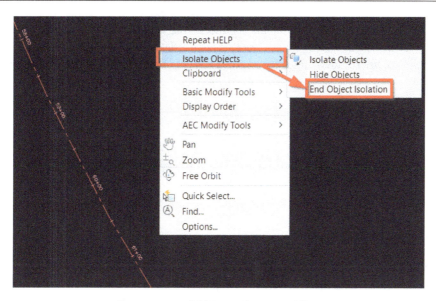

Figure 7.9 – End Object Isolation workflow

With all of our objects visible in our Survey Model file again, let's zoom out a bit to see the extents of our drawing, which should appear similar to that shown in *Figure 7.10*:

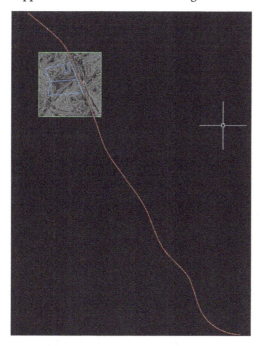

Figure 7.10 – Survey Model zoomed at extents

To ensure that we are keeping our alignments organized, we'll apply similar practices as we did with our surface creation in *Chapter 6, Surfaces - The First Foundational Component to Designs within Civil 3D*. With that, we'll use the following steps to add a level of organization in our files as it relates to our alignments:

1. Go back into the **Prospector** tab in the Toolspace.

2. Expand the **Alignments** category.

3. Right-click on **Centerline Alignments**.

4. Select the **Create Folder** option.

5. In the **Create Folder** dialog box, we'll identify the **New Folder Name** value as `Existing Conditions`.

6. Click the **OK** button.

7. Go back into the **Prospector** tab in the Toolspace.

8. Select our **ALG – Existing York Hwy – FromSurveyPoints** alignment.

9. Drag and drop our **ALG – Existing York Hwy – FromSurveyPoints** alignment into the `Existing Conditions` folder we just created (the final organization should look similar to that shown in *Figure 7.11*):

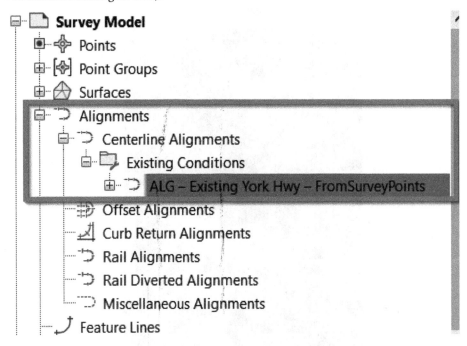

Figure 7.11 – Centerline alignment organization

The final step for us to take as it relates to our existing alignment is to create a data shortcut of our **ALG – Existing York Hwy – FromSurveyPoints** alignment so that it can be data-referenced in our design files, which will allow us to properly tie our proposed road alignment into them.

The following steps are outlined in *Chapter 3, Sharing Data within Civil 3D*, and *Chapter 5, Leveraging Points, Lines, and Curves*. We can right-click on our **Data Shortcuts** project, select **Create Data Shortcuts**, and check the box next to our **ALG – Existing York Hwy – FromSurveyPoints** alignment, as shown in *Figure 7.12*:

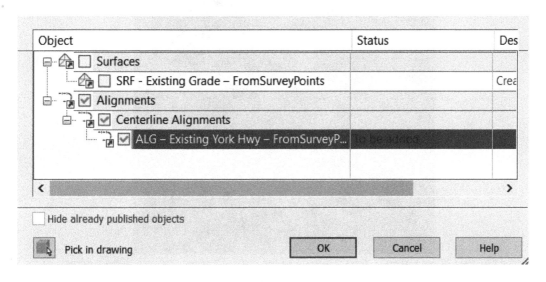

Figure 7.12 – Add ALG – Existing York Hwy – FromSurveyPoints alignment to Data Shortcuts

We're now ready to save and close out of our existing Survey Model.dwg file and move into our design files. After closing out of our Survey Model.dwg file, Civil 3D will default back to the start screen, where we will select **New | Browse templates**, as shown in *Figure 7.13*:

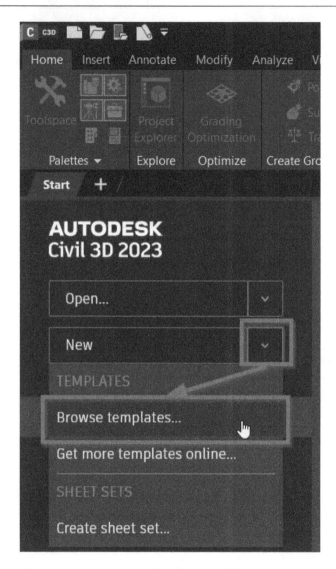

Figure 7.13 – Start screen – Creating a new file from templates

A **Select Template** dialog box will then appear, at which point we'll want to navigate to our Practical Autodesk Civil 3D 2024\Chapter 7 location, select our Company Template File. dwt file, and select **Open** in the lower right-hand corner of the **Select Template** dialog box, as shown in *Figure 7.14*:

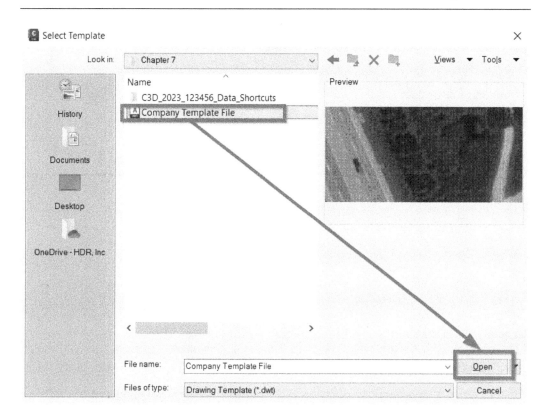

Figure 7.14 – Starting a new file with our company drawing template file

With our new file created using our `Company Template File.dwt` file, we'll follow the standard AutoCAD workflows to save our file as `Alignment Model.dwg` and to attach our `Survey Model.dwg` file as an overlay, all within our `Practical Autodesk Civil 3D 2024\ Chapter 7` location.

Next, we'll want to jump back into the **Prospector** tab in the Toolspace, and then set the **Working Folder** option of our **Data Shortcuts** project to the `Practical Autodesk Civil 3D 2024\ Chapter 7` location and select the **C3D_2024_123456_Data_Shortcuts** project, as shown in *Figure 7.15*:

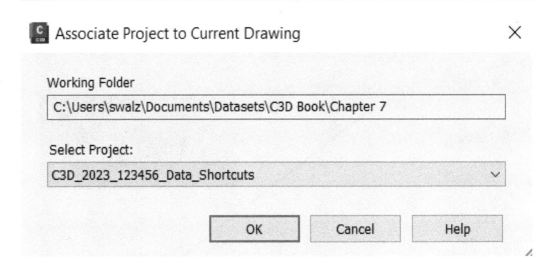

Figure 7.15 – Associate Project to Current Drawing

After our Civil 3D **Data Shortcuts** project has been associated with our current file, we can then safely create data references of our **SRF – Existing Grade – FromSurveyPoints** Surface Model and our **ALG – Existing York Hwy – FromSurveyPoints** alignment into our current `Alignment Model.dwg` file.

To do this, we need to expand the **Surfaces** and **Alignments** categories and our `Existing Conditions` folders, right-click on each element (shown in *Figure 7.16*), and select a **Create Reference** option for each:

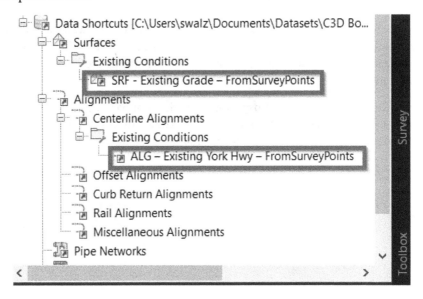

Figure 7.16 – Creating data references of identified Civil 3D objects

> **Note**
>
> Since we have our `Survey Model.dwg` file already externally referenced into our file (as an overlay), we don't necessarily need to view the Surface Model, but will need it to reference from a design standpoint a little later on down the road. That said, when we select the **Create Reference** option for our **SRF – Existing Grade – FromSurveyPoints** Surface Model, we can apply the **_No Display Surface Display** style so that it doesn't get in the way of our selections later on down the road.

With our `Alignment Model.dwg` file set up, we have all elements available to us to represent and reference the existing built environment. We are now ready to design our residential subdivision. One of the first design objects we'll want to start with is creating the main road alignment that will allow us to access our residential subdivision.

Depending on your preference, feel free to toggle off the aerial imagery shown in the background. To do this, you can go back to the **Geolocation** ribbon and select the **Map Off** option, as shown in *Figure 7.17*:

Figure 7.17 – Geolocation: Map Off

To create our **Main Residential Subdivision** alignment, we'll go up to the **Home** ribbon, the**Create design** panel, and then select the **Alignment Creation Tools** option in the **Alignment** dropdown, as shown in *Figure 7.18*):

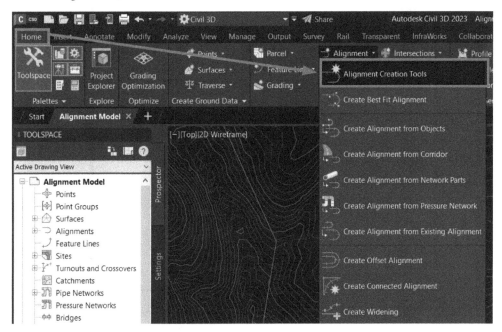

Figure 7.18 – Alignment Creation Tools

Once selected, a **Create Alignment** dialog box will appear, where we'll make the following selections, as shown in *Figure 7.19*:

- **Name**: **ALG - Subdivision Main Road - Access**
- **Type**: **Centerline**
- **Description**: **Subdivision access road from York Hwy**
- **Alignment style**: **Proposed**
- **Alignment label style**: **Major and Minor only**

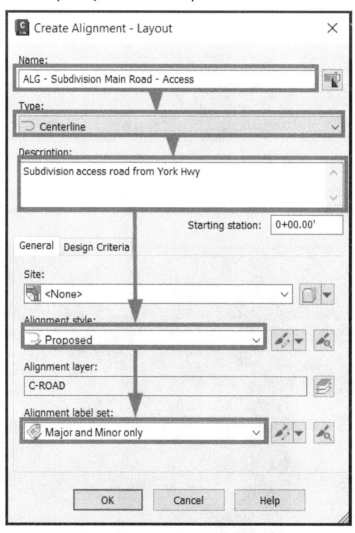

Figure 7.19 – Create Alignment dialog box

Additionally, you'll notice that in our **Create Alignment** dialog box, we have **General** and **Design Criteria** tabs. If we select the **Design Criteria** tab before selecting the **OK** button at the bottom of the **Create Alignment** dialog box, we gain access to—and have the ability to apply—design-based rules to our alignment.

That said, we'll want to make the following selections (shown in *Figure 7.20*) as well, and then we can select the **OK** button to exit out of the dialog box:

- **Starting design speed**: 25 mi/h

- **Use criteria-based design**: *Check the box*

- **Use design criteria file**: *Check the box*

- **Use design check set**: *Check the box*

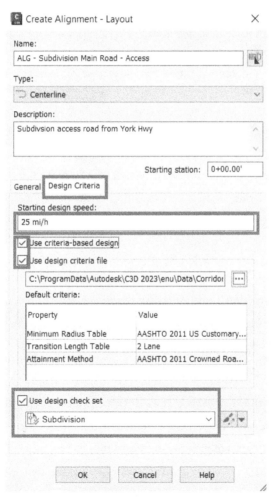

Figure 7.20 – Create Alignment dialog box – Design Criteria

Next, an **Alignment Layout Tools** toolbar will appear (similar to that shown in *Figure 7.21*). This toolbar provides all the necessary tools we could ever imagine needing during the alignment creation and editing processes:

Figure 7.21 – Alignment Layout Tools toolbar

Running through the tools as numbered in the preceding screenshot, we have the following:

1. **Continuous Geometry Creation Tools**: Allows us to lay out our alignment geometry using a **Tangent to Tangent** method. Using this method allows us to apply or disregard connecting curves automatically during the alignment creation process.

2. **Insert PI**: After our alignment creation has started, we can add new **points of intersection (PIs)** within our alignment.

3. **Delete PI**: After our alignment creation has started, we can remove PIs within our alignment.

4. **Break-Apart PI**: Allows us to define a break in our alignment geometry by identifying the nearest PI.

5. **Individual Line Creation Tools**: Allows us to individually create Fixed, Floating, or Free lines.

6. **Individual Curve Creation Tools**: Allows us to individually create Fixed, Floating, or Free curves.

7. **Floating Line with Spiral Creation Tools:** Provides two methods for creating a Floating line with spirals.

8. **Curve with Spiral Creation Tools**: Allows us to create Floating or Free curve and spiral combinations.

9. **Individual Spiral Creation Tools**: Allows us to create Fixed or Free spirals.

10. **Convert AutoCAD Line and Arc**: Allows us to convert existing lines and arcs to alignment geometry (a similar concept to the **Create Alignment From Object** method we performed in our `Survey Model.dwg` file).

11. **Reverse Sub-Entity Direction**: Allows us to reverse the direction of a Fixed line or curve.

12. **Delete Sub-Entity**: Allows us to remove a sub-entity from our alignment.

13. **Edit Best Fit Data for All Entities**: Allows us to edit regression data associated with our alignment.

14. **Pick Sub-Entity**: Allows us to pick sub-entities within our alignment to analyze.

15. **Sub-Entity Editor**: Allows us to analyze parameters associated with sub-entities within our alignment.

16. **Alignment Grid View**: Allows us to quickly view alignment geometry in table form within our Panorama.

17. **Undo**: Allows us to undo the previous command(s).

18. **Redo**: Allows us to reapply or redo commands that were undone.

Now that we have a decent idea as to which alignment creation tools we have at our disposal within Civil 3D, let's go ahead and start laying out our **ALG - Subdivision Main Road - Access** geometry using **Continuous Geometry Creation Tools** (listed as *#1* in *Figure 7.21*).

If we select the down arrow next to this icon, we'll want to make sure that the **Tangent to Tangent with Curves** option is checked. Once checked, we are ready to lay out our **ALG - Subdivision Main Road - Access** alignment.

By using the midpoint **OSNAP** (abbreviated from **Object Snap**), we'll snap to the Station **34+00** marker along our **ALG – Existing York Hwy – FromSurveyPoints** alignment, as shown in *Figure 7.22*. This will essentially mark the intersection and beginning of our **ALG - Subdivision Main Road - Access** alignment:

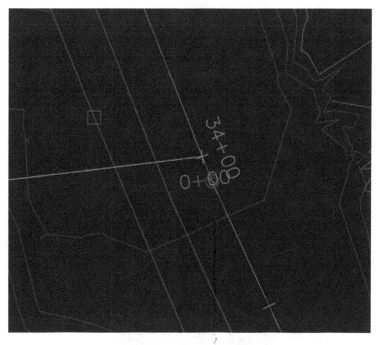

Figure 7.22 – ALG - Subdivision Main Road – Access intersection point
with ALG – Existing York Hwy – FromSurveyPoints

Next, we'll click three more times in different locations within our site to define all four of our PI locations (these locations are indicated with a red circle in *Figure 7.23*):

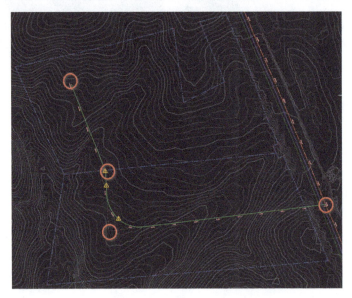

Figure 7.23 – ALG - Subdivision Main Road – Access PIs

Finally, using the same steps we used to create our **ALG - Subdivision Main Road - Access** alignment, we will now go ahead and create one more alignment to represent the centerline of our subdivision side road.

This subdivision side road will be called **ALG - Subdivision Side Road - Cul-De-Sac** and will apply the exact same general and design parameters as was assigned to our **ALG - Subdivision Main Road - Access** alignment.

As we lay our **ALG - Subdivision Side Road - Cul-De-Sac** alignment out, we'll want to snap to the **15+00** station marker along our **ALG - Subdivision Main Road - Access** alignment and place three additional PIs, as indicated in *Figure 7.24*:

Figure 7.24 – ALG – Subdivision Side Road – Access PIs

With our **Existing Surface Model** and our **Proposed Alignment** geometry created, we have officially set ourselves up with a great foundation to continue developing and building up our design. These key foundational items provide the necessary direction and are a base for us to build our design on top of, ensuring that we are fully integrating our **Proposed Residential** subdivision into the existing built environment. As we progress through the remainder of the chapters, we'll continue to realize how important these two foundational objects are to Civil 3D designs.

With this understanding, next, we will dive into the nuances of alignments when it comes to styles, manipulation, and design validation.

Understanding alignment styles

Understanding what we're looking at and the various alignment styles that we have in our arsenal is an important step to fully realizing the benefits of alignment geometry and generating these objects in a dynamic nature. In this section, we'll familiarize ourselves with the various alignment styles available to us and learn how to display these styles in different circumstances.

For this section, we'll continue working in our `Alignment Model.dwg` file, located in our `Practical Autodesk Civil 3D 2024\Chapter 7` subfolder. Next, we'll go over to the Toolspace, select the **Settings** tab, and then expand the **Alignment** category by clicking the + icon next to **Alignment**.

We'll notice that we have the following list of subcategories associated with alignments (also displayed in *Figure 7.25*):

- **Alignment Styles**: Here, we will be able to view our comprehensive list of available display styles associated with alignment geometry.

- **Design Checks**: Here, we will be able to specify additional design parameters that need to be applied to our alignment lines, curves, spirals, and tangent intersections. As we set these parameters here, we can add to our design check sets for a fully comprehensive check of individual sub-entities that make up our alignment.

- **Label Styles**: Here, we will be able to view our comprehensive list of available label styles that can be applied to our alignment geometry for stationing and annotation purposes; these can also display specified symbology.

- **Table Styles**: Here, we will be able to view our comprehensive list of available table styles that can be displayed for sheeting purposes as they relate to our alignment geometry.

- **Commands**: Here, we will be able to view a comprehensive list of available commands that can simply be typed into our command line to perform specific tasks associated with alignments:

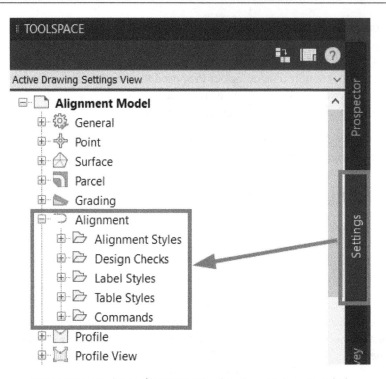

Figure 7.25 – Toolspace | Settings tab: Alignment Styles expanded

Starting with **Alignment Styles**, if we click on the + icon next to it, we will then see that comprehensive list of alignment styles that reside in our current drawing.

> **Note**
>
> These are also contained within our Company Template File.dwt drawing template that we created as a starting point for new files. That said, all styles listed from here on will be included in any new file, provided you are using the Company Template File.dwt as a base.

Types of alignment styles

Once expanded, we'll see the following list of alignment styles that can be applied to our Alignment objects within our current file for display purposes (also refer to *Figure 7.26*):

- **Basic**: Applying this display to our alignment will essentially inherit the layer properties that the alignment has been assigned to

- **Existing**: Applying this display to our alignment allows us to display it as a background object during the plotting process

- **Intersection Basic**: Applying this display to our alignment will typically be utilized during intersection designs

- **Layout**: Applying this display to our alignment will place lines, curves, and spirals on separate layers with different color assignments that will allow us to quickly visually locate each type of sub-entity

- **Offsets**: Applying this display to our alignment will typically be utilized as we generate offset alignments to project, or target, our corridor subassemblies to

- **Proposed**: Applying this display to our alignment allows us to display as a proposed, or bold, object during the plotting process:

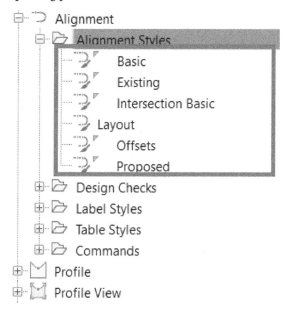

Figure 7.26 – Alignment styles

Alignment design checks

Next up are our design checks that we can apply during alignment creation and design validation processes. If we click on the + icon next to **Design Checks**, we will then see a comprehensive list of design checks that reside in our current drawing.

Once expanded, we'll see the following list of design checks that can be applied to our Alignment objects within our current file (also refer to *Figure 7.27*):

- **Design Check Sets**: Allows us to apply multiple design parameters to our Alignment objects

- **Line**: Allows us to define individual design parameters associated with lines specifically

- **Curve**: Allows us to define individual design parameters associated with curves specifically

- **Spiral**: Allows us to define individual design parameters associated with spirals specifically

- **Superelevation**: Allows us to define individual design parameters associated with superelevations specifically

- **Tangent Intersection**: Allows us to define individual design parameters associated with tangent intersections specifically:

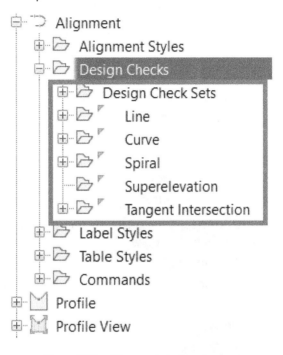

Figure 7.27 – Alignment design checks

Alignment label styles

Next, we'll visit our alignment label styles. With that, we'll go ahead and click on the + icon next to **Label Styles**, at which point we'll then see a comprehensive list of alignment label styles that reside in our current drawing.

Once expanded, we'll see the following list of label styles that can be applied to our alignments within our current file for annotation display purposes (also displayed in *Figure 7.28*):

- **Label Sets**: Allows us to apply multiple label styles to our alignments as a comprehensive set

- **Station**: Allows us to create and apply individual station labels to our Alignment objects

- **Station Offset**: Allows us to create and apply individual station offset labels to our Alignment objects

- **Line**: Allows us to create and apply individual line labels to our Alignment objects

- **Curve**: Allows us to create and apply individual curve labels to our Alignment objects

- **Spiral**: Allows us to create and apply individual spiral labels to our Alignment objects

- **Tangent Intersection**: Allows us to create and apply individual tangent intersection labels to our Alignment objects

- **Point of Intersection**: Allows us to create and apply individual PI labels to our Alignment objects

- **Cant Information**: Allows us to create and apply individual cant labels to our Alignment objects:

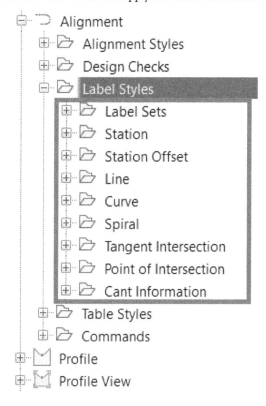

Figure 7.28 – Alignment label styles

To add alignment labels, navigate to the **Annotate** tab, then click the drop-down arrow on the **Add Labels** button. Then, navigate to **Alignment**, and you will see all the options you have for creating Alignment Labels:

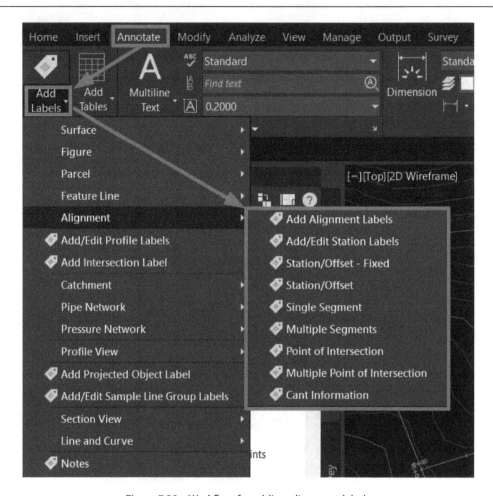

Figure 7.29 – Workflow for adding alignment labels

Alignment table styles

Next, we'll visit our Alignment table styles. With that, we'll go ahead and click on the + icon next to **Table Styles**, at which point we'll then see a comprehensive list of Alignment table styles that reside in our current drawing.

Once expanded, we'll see the following list of table styles (also refer to *Figure 7.30* for a detailed breakdown of Alignment table styles in our current file) that can be added to our drawings for sheeting purposes:

- **Line:** This table style will be dynamically linked to our alignment geometry and will display specified information related to the lines contained within

- **Curve:** This table style will be dynamically linked to our alignment geometry and will display specified information related to the curves contained within

- **Spiral**: This table style will be dynamically linked to our alignment geometry and will display specified information related to the spirals contained within

- **Segment**: This table style will be dynamically linked to our alignment geometry and will display specified information related to all segments contained within:

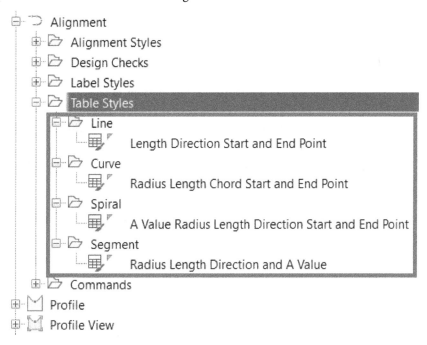

Figure 7.30 – Alignment table styles (detailed version)

With that overview of all applicable styles that can be applied to our alignments out of the way, let's get a better understanding of what some of these actually mean and represent, how we can apply different styles in different circumstances and scenarios, and how we can enhance our overall understanding of alignments and the impact they can have on our designs.

Alignment manipulation and management

In a majority of cases, we're going to want to maintain the display of our alignments as we create them. Occasionally, we'll come across a need to make some adjustments to how we are displaying our alignment, station referencing, how we are manipulating our alignment geometry, and maybe even applying a different set of design parameters to our Alignment objects.

To access any of these options, we can simply select our Alignment object in Civil 3D, right-click in **Model Space**, and select the **Alignment Properties** option. Once selected, we'll be presented with our **Alignment Properties** dialog box.

Along the top of our **Alignment Properties** dialog box, we have the following series of tabs, where we can access these manipulation options mentioned, as shown in *Figure 7.31*:

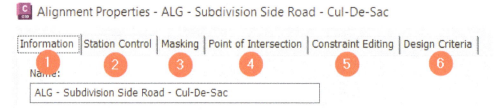

Figure 7.31 – Alignment Properties: manipulation options

- **Information**: Allows us to make changes to our alignment name, description, and styles

- **Station Control**: Allows us to make changes to our station information and the starting point of our alignment and add station equations if needed

- **Masking**: Allows us to hide specific station ranges along our alignment if needed

- **Point of Intersection**: Allows us to change the display of PIs if not implied by fixed tangent intersections

- **Constraint Editing**: Allows us to change the way our tangency and parameter constraints function while performing edits to our alignment

- **Design Criteria**: Allows us to add different parameters to our alignment for design check purposes

In addition to the manipulation options available to us in the **Alignment Properties** dialog box, if we were to simply select our Alignment object in our Model Space, we would then be able to view the **Context** ribbon along the top of our screen, as shown in *Figure 7.32*:

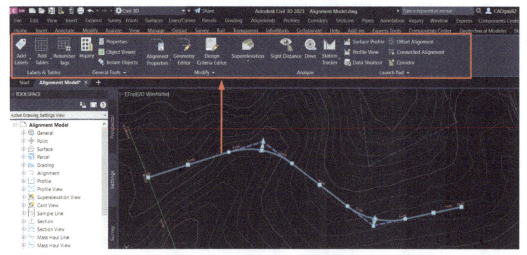

Figure 7.32 – Alignment Context ribbon

The **Contextual** ribbon is a new set of options that will display in ribbon form and will allow us to perform additional edits, inquiries, and analyses on our selected geometry. In this case, we have selected our alignment, in which an **Alignment Contextual** ribbon appears that allows us to further interrogate, manipulate, and analyze our **ALG - Subdivision Side Road - Cul-De-Sac** alignment.

Running from left to right as shown in *Figure 7.32*, we have the following panels with tools available to us:

- **Labels & Tables**: Grants us access to various annotative tools
- **General Tools**: Grants us access to various inquiry tools that will display information associated with our selected alignment
- **Modify**: Grants us access to various tools that allow us to manipulate our alignment geometry
- **Analyze**: Grants us access to various analysis tools to further examine our alignment geometry
- **Launch Pad**: Grants us access to major tools/workflows that will enable further development of our design and collaboration

> Note
>
> Analysis tools available in our **Alignment Contextual** ribbon will require a profile to be created along our alignment first.

As we progress through our design, we will revisit many of these tools and capabilities available to us to further validate our alignments and design in general. The unfortunate reality at this point in our design progression is that we just don't have enough design objects created at the moment to truly take advantage of these tools and capabilities available to us in the **Contextual** ribbon.

Summary

With that, although we have performed a bit of discovery as it relates to uncovering how big a part alignments play in our design, there are many additional components that we need to dynamically link to alignment geometry to truly appreciate and realize the value they bring to the table.

That said, although we have two foundational Civil 3D objects created, we need to continue progressing further to tie our third foundation Civil 3D object into the mix, which will really bring to light how critical these steps are at the beginning of our designs.

In our next chapter, we'll begin exploring Profiles that will be referenced and tied to our first two foundational objects (Surfaces and Alignments). Once we're able to form a dynamic link between all three of these foundational objects, we can begin exploring additional design and analysis tools that are not currently available to us.

8

Profiles - The Third Foundational Component to Designs within Civil 3D

As discussed in the past few chapters, surfaces, alignments, and profiles are often considered to be the three major foundational components required to build a true civil infrastructure design model within Civil 3D. In *Chapter 6, Surfaces - The First Foundational Component to Designs within Civil 3D*, we took a deep dive into surfaces, while in *Chapter 7, Alignments - The Second Foundational Component to Designs within Civil 3D*, we took a deep dive into alignments.

Along the way, we've learned several ways to generate, modify, and analyze both surface models and alignment geometry. We realized towards the end of *Chapter 7* that from an analysis standpoint, we fall slightly short with some of the functionality available to us, as we have yet to generate profiles that tie the Surfaces and Alignments together.

In this chapter, we'll take a deep dive into the world of profiles, where we'll begin to realize how key this particular component is to establish the foundation for our Residential Subdivision design. Once our profiles are created, we'll also revisit some of the missing analysis capabilities that we were unable to perform in *Chapter 6, Surfaces - The First Foundational Component to Designs within Civil 3D* as they are related to alignments.

With that in mind, in this chapter, we'll focus on the following key aspects as they relate to profiles in Civil 3D:

- Understanding ways to create a profile
- Setting up our profile views
- Creating design profiles
- Understanding profiles and profile view styles
- Further analyzing profile and alignment geometry

As previously discovered with surfaces and alignments, we have begun to realize how these modeled objects play a big part in setting the foundation for our design. With profiles being the missing link to tie all these types of objects together, let's not waste any more time and jump right into it.

Let's go ahead and open up Civil 3D, or go to the **Start** screen if already open, and create a new drawing using similar steps to those outlined in *Chapter 7, Alignments - The Second Foundational Component to Designs within Civil 3D*. We can use the Company Template File.dwt file located in Practical Autodesk Civil 3D 2024\Chapter 8 and select **Open** in the lower right-hand corner of the **Select Template** dialog box.

Once our new file is created, we'll want to save it as the Grading Model.dwg file to the Practical Autodesk Civil 3D 2024\Chapter 8\Model location.

As discussed back in *Chapter 3, Sharing Data within Civil 3D* roadway centerline alignments and their profiles are often used to create corridor objects, which will ultimately generate our proposed surface model (or at least a portion of it). Therefore, we will need the grading model to contain the profiles that will display the existing and proposed vertical geometry.

Next, we want to attach the Survey Model.dwg and Alignment Model.dwg files as an overlay, all contained within the Practical Autodesk Civil 3D 2024\Chapter 8\Model location. Then we want to jump back into the **Prospector** tab in **TOOLSPACE**, and then update the **Working Folder** field of the **Data Shortcuts** project to the Practical Autodesk Civil 3D 2024\Chapter 8 location and select the **C3D_2024_123456_Data_Shortcuts** project, as shown in *Figure 8.1*.

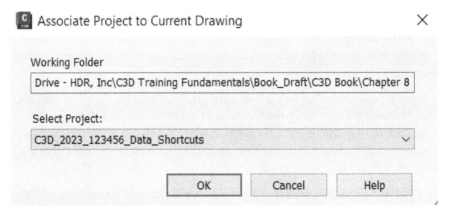

Figure 8.1 – Associate Project to Current Drawing

After our Civil 3D Data Shortcuts project has been associated with the current file, we can then safely create the data references for the **SRF – Existing Grade – FromSurveyPoints** surface model, along with the **ALG – Existing York Hwy – FromSurveyPoints**, **ALG - Subdivision Main Road - Access**, and **ALG - Subdivision Side Road - Cul-De-Sac** alignments and add them to our current Grading Model.dwg file.

To do this, we need to expand our surfaces and alignments categories and the `Existing Conditions` folders by right-clicking on each element, as shown in *Figure 8.2*) and selecting the **Create Reference** option for each.

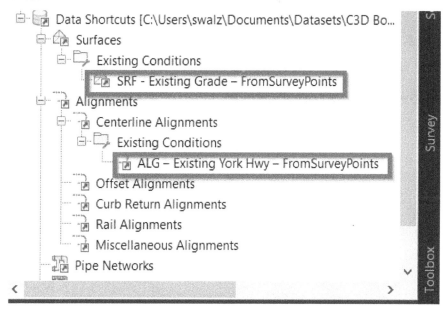

Figure 8.2 – Create data references for identified Civil 3D objects

Then, with the `Grading Model.dwg` file set up for us to include all elements we have available to us to represent and reference the existing built environment, along with the beginnings of the future built environment, we are now ready to begin designing our residential subdivision layout.

Technical requirements

The exercise files for this chapter are available at `https://packt.link/UoiPn`

Understanding the ways to create a profile

In this section, we'll begin exploring multiple ways to create a profile, all while getting a full understanding of the various profile creation tools available to us within the Civil 3D environment.

To access our profile creation tools, we need to activate the **Home** ribbon along the top of our Civil 3D session, go to the **Create Design** panel, and click on the down arrow next to where it says **Profile**.

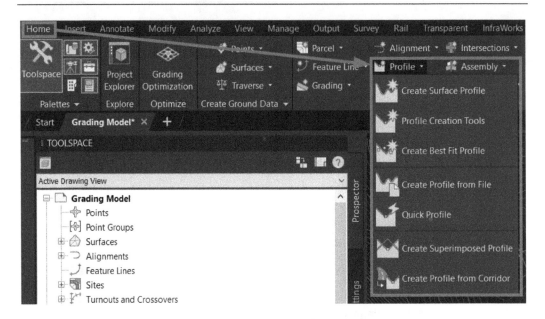

Figure 8.3 – Profile creation tools

As shown in *Figure 8.3*, we have the following profile creation tools available to us:

- **Create Surface Profile**: This allows us to create a profile referencing an alignment and surface model existing in the current file

- **Profile Creation Tools**: This allows us to create vertical geometry (also known as a profile) along an existing horizontal alignment (a profile view must already be created within the current file)

- **Create Best Fit Profile**: This allows us to create optimal vertical geometry (a profile) along an existing horizontal alignment based on various parameters identified (a profile view must already be created within the current file)

- **Create Profile from File**: This allows us to create a profile by importing an ASCII survey point file

- **Quick Profile**: Allows us to create a temporary profile and profile view using various 2D and 3D objects, including 2D/3D polylines, feature lines, and points

- **Create Superimposed Profile**: Allows us to project profiles into other profile views

- **Create Profile from Corridor**: Allows us to create a profile from a feature line generated/extracted from a corridor model

Let's now go ahead and create the first profile for the **ALG - Subdivision Main Road – Access** alignment. To do so, we need to use the **Create Surface Profile** command (refer to *Figure 8.3* for the location of this tool).

Once the **Create Profile from Surface** dialog box has been activated, we need to take the following steps to connect the proposed horizontal alignment with the existing surface model, as shown in *Figure 8.4*:

1. In the **Alignment** field, select the **ALG - Subdivision Main Road – Access** alignment.

2. In the **Select Surfaces** field, select the **SRF – Existing Grade – FromSurveyPoints** surface model.

3. Once both the alignment and surface have been selected, click on the **Add>>** button, at which point we will see **Profile list** populate with the corresponding alignment and surface.

4. Click on the **OK** button to exit.

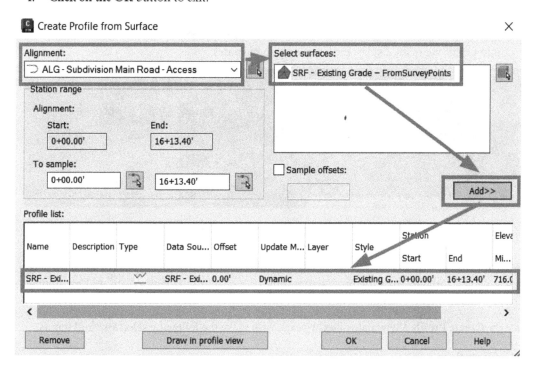

Figure 8.4 – Create Profile from Surface dialog box

Setting up profile views

Now that we've connected the **ALG - Subdivision Main Road – Access** alignment to the **SRF – Existing Grade – FromSurveyPoints** surface model, we can now create the profile view in preparation for developing the proposed grading model.

By going back to the **Home** ribbon and then proceeding to the **Profile** and **Section Views** panels, we can access our profile view tools. If we click on the down arrow next to where it says **Profile View**, we'll see a drop-down menu appear with the following options, shown in *Figure 8.5*:

- **Create Profile View**: This allows us to create a singular profile view along with an alignment

- **Create Multiple Profile Views**: This allows us to create multiple profile views in the event that a particular alignment is long (typically used for sheeting purposes)

- **Project Objects to Profile View**: This allows us to project objects into profile views for display and coordination purposes

- **Add Crossings to Profile View**: This allows us to project linear objects that cross alignments into profile views for display and coordination purposes

Figure 8.5 – Profile View tools

For the time being, we want to select the **Create Profile View** option. Being that we're not at the stage of sheeting yet, and only focusing on design, profiles are easier to manage, create, and manipulate when they are all displayed in one continuous profile view.

Once the **Create Profile View** dialog box appears, we'll make the following selections and inputs:

- **General**:

    - **Select alignment: ALG - Subdivision Main Road - Access**

    - **Profile view name: PRV - Subdivision Main Road - Access**

    - **Description: Profile View along 'Subdivision Main Road - Access' Alignment**

    - **Profile view style: Land Desktop Profile View**

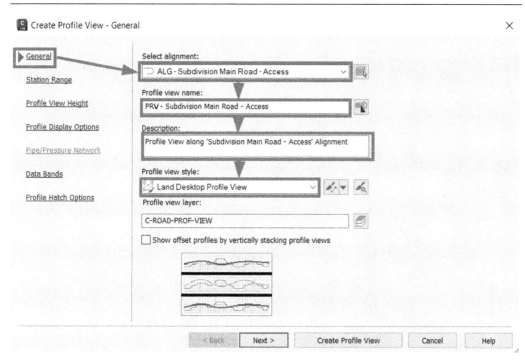

Figure 8.6 – Create Profile View – General

- **Station Range**: **Automatic**

Figure 8.7 – Create Profile View – Station Range

- **Profile View Height**: **Automatic**

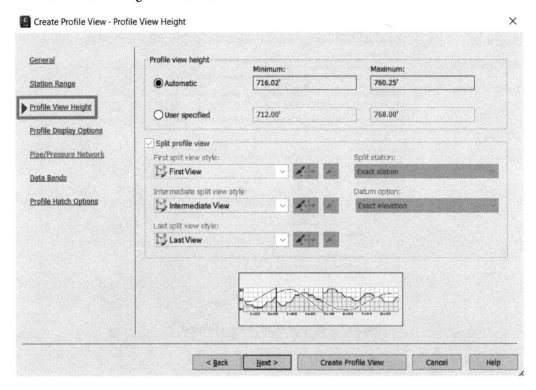

Figure 8.8 – Create Profile View – Profile View Height

- **Profile Display Options**:

    Style: **Existing Ground Profile**

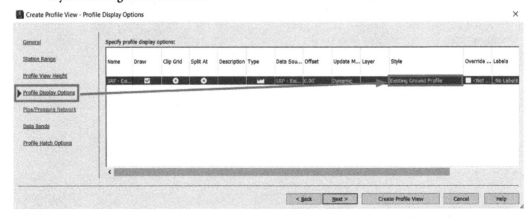

Figure 8.9 – Create Profile View – Profile Display Options

- **Data Bands:**

  Select band set: **EG-FG Elevations and Stations**

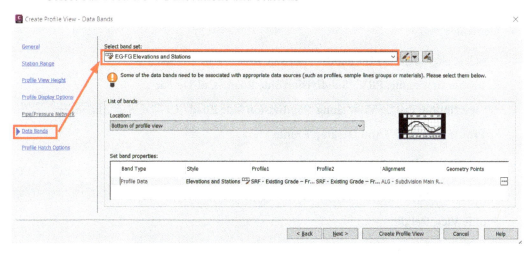

Figure 8.10 – Create Profile View – Data Bands

Finally, we select the **Create Profile View** button in the lower section of the **Create Profile View** dialog box. Once selected, we'll be prompted to define a location in which we want to place our new profile view.

Let's zoom out a little bit so that we can see the full site and left-click the mouse button to place our profile view on the right side of the site, as shown in *Figure 8.11*.

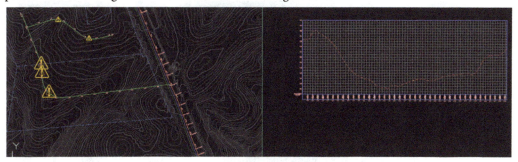

Figure 8.11 – Placing our new profile view

Using the same steps for creating a profile and profile view that we previously went through, let's go ahead and create another profile view for the **ALG - Subdivision Side Road – Cul-De-Sac** alignment.

First, create a profile of the existing surface linked to the **ALG - Subdivision Side Road-Cul-De-Sac** alignment, then use the following criteria in the **Create Profile View** dialog box (with the final output as shown in *Figure 8.12*):

- **General**:
  - **Select alignment: ALG - Subdivision Side Road – Cul-De-Sac**
  - **Profile view name: PRV - Subdivision Side Road - Cul-De-Sac**
  - **Description: Profile View along 'Subdivision Side Road - Cul-De-Sac'** Alignment
  - **Profile view style: Land Desktop Profile View**

- **Profile Station Range**:
  - **Station Range: Automatic**

- **Profile View Height**:
  - **Profile View Height: Automatic**

- **Profile Display Options**:
  - **Style: Existing Ground Profile**

- **Data Bands**:
  - **Select Band Set: EG-FG Elevations and Stations**

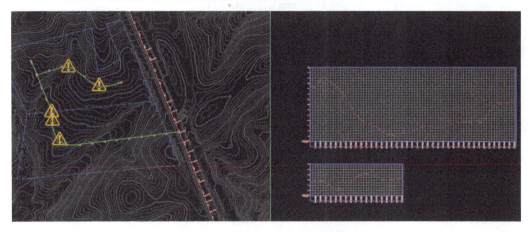

Figure 8.12 – Placing our second profile view

With our profile view created now showing our existing profile grade lines, we can now begin exploring ways we can use our profile views to start generating our proposed design grade lines and familiarizing ourselves with the design profile creation tools.

Creating design profiles

Moving over to the design side of our profile creation process, we'll need to go back up to the **Home** ribbon and the **Create Design** panel to access **Profile Creation Tools** (refer to *Figure 8.13*).

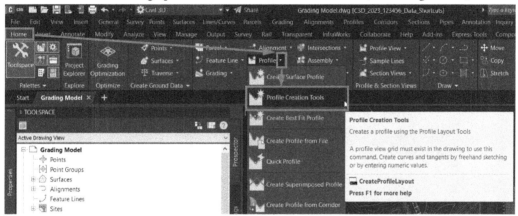

Figure 8.13 – Accessing Profile Creation Tools

Once **Profile Creation Tools** has been selected, we'll be prompted to select a profile view with which we'd like to place or associate our design profile.

Let's go ahead and select the **PRV - Subdivision Main Road – Access** profile view (the profile view we placed on top), at which point the **Create Profile - Draw New** dialog box will appear. In the **Create Profile - Draw New** dialog box, we need to fill out the fields as follows (also displayed in *Figure 8.14*):

- **Name**: **PRF - Subdivision Main Road - Access**
- **Description**: **Profile along 'Subdivision Main Road - Access' Alignment**
- **Profile style**: **Design Profile**
- **Profile label set**: **Complete Label Set**

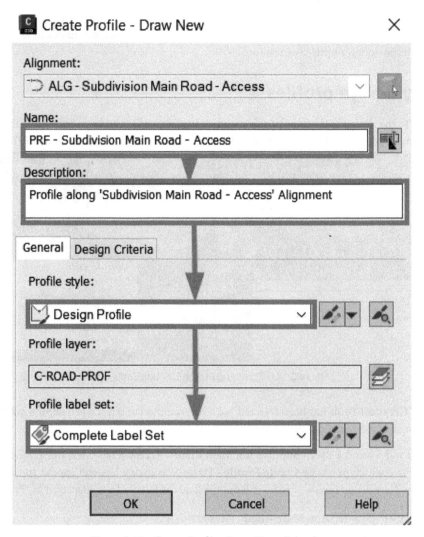

Figure 8.14 – Create Profile - Draw New dialog box

Before closing out of the dialog box, we also need to switch over to the **Design Criteria** section and use the following selections (also displayed in *Figure 8.15*):

- **Use criteria-based design**: Check the box

- **Use design criteria file**: Check the box

- **Use design check set**: Check the box and select the **Subdivision** design check set

Figure 8.15 – Create Profile – Draw New Design Criteria

After all the selections have been made, click the **OK** button at the bottom of the **Create Profile - Draw New** dialog box. The **Profile Layout Tools** toolbar will then appear, as shown in *Figure 8.16*.

This toolbar provides all the necessary tools we could ever imagine needing during the profile creation and editing processes.

Figure 8.16 – Profile Layout Tools toolbar

Running through the tools, as numbered in *Figure 8.16*, we have the following:

1. **Continuous Geometry Creation Tools**: Allow us to lay out the profile geometry using a **Tangent to Tangent** method. Using this method allows us to apply or disregard connecting curves automatically during the alignment creation process.

2. **Insert PVI**: After the profile creation has started, we can add new **Point of Vertical Intersections (PVIs)** within the profile.

3. **Delete PVI**: After the profile creation has started, we can remove PVIs within the profile.

4. **Move PVI**: After the profile creation has started, we can move PVIs within the profile.

5. **Individual Tangent Creation Tools**: Allow us to individually create fixed, floating, or free lines.

6. **Individual Vertical Curve Creation Tools**: Allow us to individually create fixed, floating, or free vertical curves.

7. **Convert AutoCAD Line and Spline**: Allows us to convert existing lines and splines to profile geometry.

8. **Insert PVIs - Tabular**: Allows us to create various types of vertical curves based on PVIs.

9. **Raise/Lower PVIs**: Allows us to raise and/or lower PVIs.

10. **Copy Profile**: Allows us to copy profiles and profile data to a specified profile view.

11. **Change Profile Design Methods**: Allows us to change our method of profile creation based on PVIs or entities.

12. **Select PVI**: Allows us to select the location of a PVI and update the parameters as required.

13. **Extend Entity**: Allows us to extend profile entities as required.

14. **Delete Subentity**: Allows us to remove a profile subentity as required.

15. **Edit Best Fit Data for All Entities**: When using the **Create Best Fit Profile** method, this tool will allow us to edit and synchronize regression data associated with the profile.

16. **Profile Layout Parameters**: Allows us to update the profile parameters as required.

17. **Profile Grid View**: Allows us to update the profile criteria in our panorama.

18. **Undo**: Allows us to undo the previous command(s).

19. **Redo**: Allows us to reapply or redo commands that were undone.

Now that we have a decent idea as to what profile creation tools we have at our disposal within Civil 3D, let's go ahead and start laying out the **PRF - Subdivision Main Road - Access** geometry using **Continuous Geometry Creation Tools** (listed as #1 in *Figure 8.16*).

If we select the down arrow next to **Continuous Geometry Creation Tools** icon, we'll want to make sure that the **Draw Tangents with Curves** option is checked. Once this is done, we are ready to lay out the **PRF - Subdivision Main Road - Access** profile.

To begin our profile design, we'll want to left-click the mouse button at the beginning of our profile where the existing grade is.

To do this, we also want to be sure that we are using OSNAPs to snap to the endpoint of our existing grade profile, as shown in *Figure 8.17*.

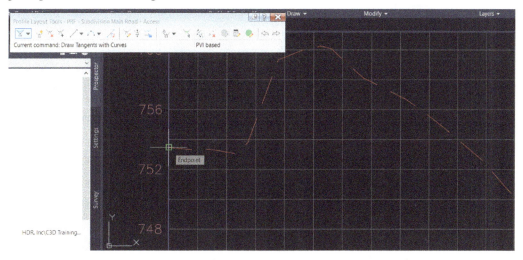

Figure 8.17 – Snap to endpoint of existing profile grade line to create our design profile

After clicking on the starting point, we'll place five more points, two of which will be approximated for the time being. The second PVI will be placed at around Station 1+00 at an elevation of around 753. The third PVI will be placed at around Station 6+00 at an elevation of around 720. The fourth PVI will be placed at around Station 12+00 at an elevation of around 728. The fourth PVI will be placed at around Station 14+80 at an elevation of around 746. And the final PVI will be placed by snapping to the endpoint of the existing grade line.

The final design profile should look similar to that shown in *Figure 8.18*.

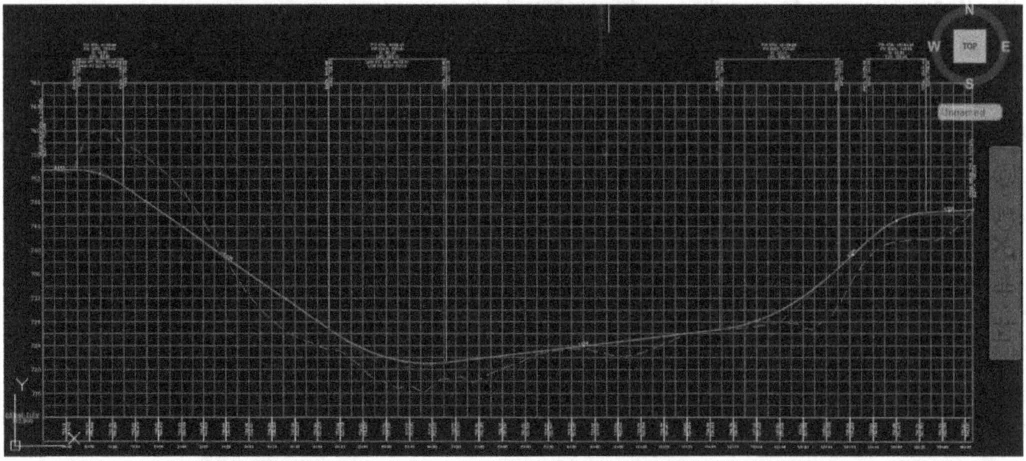

Figure 8.18 – PRF - Subdivision Main Road – Access profile view

Next, we'll want to close out of the **Profile Layout Tools** toolbar and begin the whole process again of creating our design profile inside the **PRV - Subdivision Side Road - Cul-De-Sac** profile view using the **Profile Creation Tools** method again.

We'll go ahead and fill out the fields as follows (also shown in *Figure 8.19*):

- **Name: PRF - Subdivision Side Road – Cul-De-Sac**
- **Description: Profile along 'Subdivision Side Road – Cul-De-Sac' Alignment**
- **Profile style: Design Profile**
- **Profile label set: Complete Label Set**

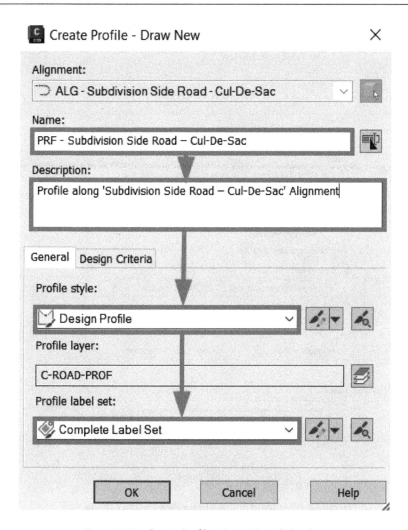

Figure 8.19 – Create Profile – Draw New dialog box

Before closing out of the **Create Profile – Draw New** dialog box, we also want to switch over to the **Design Criteria** section and use the following selections, as shown in *Figure 8.20*:

- **Use criteria-based design**: *Check the box*
- **Use design criteria file**: *Check the box*

- **Use design check set**: Check the box and select the **Subdivision** design check set

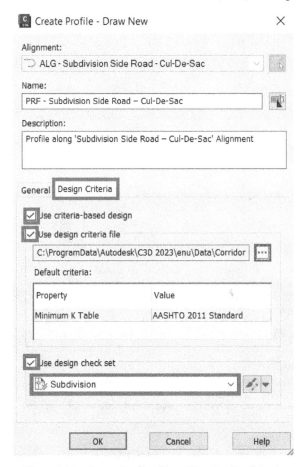

Figure 8.20 – Create Profile - Draw New Design Criteria

After all the fields have been filled out and the design criteria has been defined, click the **OK** button at the bottom of the **Create Profile - Draw New** dialog box, where the **Profile Layout Tools** toolbar will display again.

An important thing to note is that the current profile we are attempting to create or edit is always displayed in the title of the **Profile Layout Tools** toolbar (refer to *Figure 8.21*):

Figure 8.21 – Profile Layout Tools toolbar

Using the **Continuous Geometry Creation Tools** again, we want to start laying out the new **PRF - Subdivision Side Road – Cul-De-Sac** profile. Just as we did before, if we select the down arrow next to the **Continuous Geometry Creation Tools** icon, we'll want to make sure that the **Draw Tangents with Curves** option is checked before beginning to lay out the profile geometry.

To begin our profile design, we'll want to left-click the mouse button at the beginning of the profile where the existing grade is, using the endpoint ONSAP again. After clicking on the starting point, we'll place three more points, two of which will be approximated for the time being.

The second PVI will be placed at around Station 2+00 at an elevation of around 740. The third PVI will be placed at around Station 4+40 at an elevation of around 746. And the final PVI will be placed by snapping to the endpoint of the existing grade line. The final design profile should look similar to that shown in *Figure 8.22*.

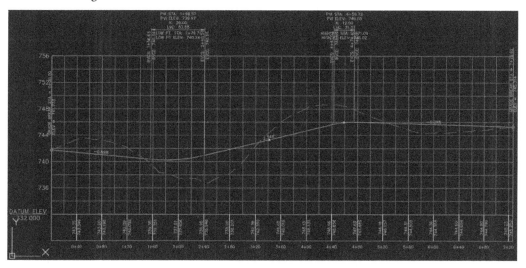

Figure 8.22 – PRF - Subdivision Side Road – Cul-De-Sac profile view

Now we have begun to establish some basic proposed elevations that we'd like to apply to our residential subdivision roads.

Let's dive into the various styles and settings associated with profiles and profile views to better understand how best we can manipulate, manage, analyze, and prepare these Civil 3D objects for design and sheeting purposes.

Understanding profiles and profile view styles

For this section, we'll continue working in the Grading Model.dwg file, located in the Practical Autodesk Civil 3D 2024\Chapter 8 subfolder. Next, we'll go over to **TOOLSPACE**, select the **Settings** tab, and then expand the **Profile** category by clicking the + icon next to **Profile**.

Notice that we have the following list of subcategories associated with profiles, as shown in *Figure 8.23*:

- **Profile Styles**: Here, we can view the comprehensive list of available display styles associated with profile geometry.

- **Design Checks**: Here, we are able to specify additional design parameters that need to be applied to our profile lines and curves. As we set these parameters here, we can add them to **Design Check Sets** for a fully comprehensive check of individual sub-entities that make up our profile.

- **Label Styles**: Here, we can view a comprehensive list of available label styles that can be applied to the profile geometry for stationing and annotation purposes; these can also display specified symbology.

- **Commands**: Here, we are able to view a comprehensive list of available commands that can simply be typed into the command line to perform specific tasks associated with profiles.

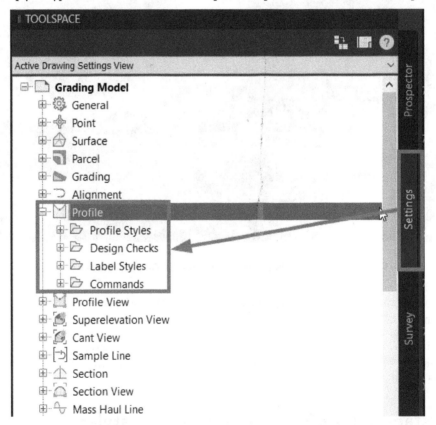

Figure 8.23 – TOOLSPACE | Settings tab – Profile expanded

Starting with profile styles, if we click on the + icon next to **Profile Styles**, we will then see a comprehensive list of profile styles that reside in our current drawing.

> **Note**
>
> These are also contained within the `Company Template File.dwt` drawing template we created as a starting point for new files. That said, all styles listed here-on-out will be included in any new file, provided you are using the `Company Template File.dwt` file as that base.

Profile Styles

Once expanded, we'll see the following list of profile styles that can be applied to the profile objects within our current file for display purposes (also refer to *Figure 8.24*):

- **Basic**: Applying this display to our profile means it will essentially inherit the layer properties that the alignment has been assigned to

- **Design Profile**: Applying this display to our profiles allows us to display as proposed in our design files

- **Existing Ground Profile**: Applying this display to our profiles allows us to display as existing in our design files

- **Intersection Basic**: Applying this display to our profile is typically utilized during intersection designs

- **Layout**: Applying this display to our profile will place lines and curves on separate layers with different color assignments allowing us to quickly visually locate each type of sub-entity

- **Left Sample Profile**: Applying this display to our profile allows us to quickly distinguish left offset lines

- **Right Sample Profile**: Applying this display to our profile allows us to quickly distinguish right offset lines

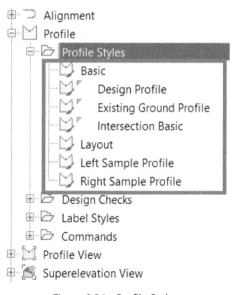

Figure 8.24 – Profile Styles

Design Checks

Next up are the design checks that we can apply during profile creation and design validation processes. If we click on the + icon next to **Design Checks**, we will then see a comprehensive list of design checks that reside in our current drawing.

Once the list is expanded, we'll see the following list of design checks that can be applied to our profile objects within our current file (also refer to *Figure 8.25*):

- **Design Check Sets**: This allows us to apply multiple design parameters to our profile objects
- **Line**: Allows us to define individual design parameters associated with lines specifically
- **Curve**: Allows us to define individual design parameters associated with curves specifically

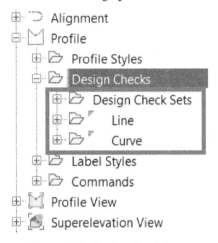

Figure 8.25 – Design Check Sets

Next, we'll visit profile label styles. Go ahead and click on the + icon next to **Label Styles**, at which point we'll see a comprehensive list of profile label styles that reside in our current drawing.

Label Styles

Once expanded, we'll see the following list of label styles that can be applied to our profiles within our current file for annotation display purposes, as shown in *Figure 8.26*:

- **Label Sets**: This allows us to apply multiple label styles to our profiles as a comprehensive set
- **Station**: Allows us to create and apply individual station labels to our profile objects
- **Grade Breaks**: Allows us to create and apply individual grade break labels to our profile objects

- **Line**: Allows us to create and apply individual line labels to our profile objects

- **Curve**: Allows us to create and apply individual curve labels to our profile objects

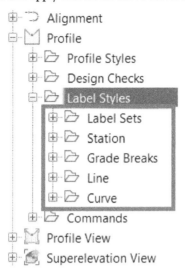

Figure 8.26 – Profile | Label Styles

Next, we'll go over to **TOOLSPACE**, select the **Settings** tab, and then expand the **Profile View** category by clicking the + icon next to **Profile View**.

Profile View

Notice that we have the following list of subcategories associated with profile views (displayed in *Figure 8.27*):

- **Profile View Styles**: Allows us to apply and view various display options that control grids/graphs and standard annotation/labels applied throughout our profile views

- **Label Styles**: Allows us to specify the appearance of individual labels placed throughout our profile views

- **Band Styles**: Allows us to specify the appearance and data to be displayed along the top or bottom of our profile views

- **Commands**: Here, we can view a comprehensive list of available commands that can simply be typed into the command line to perform specific tasks associated with profile views

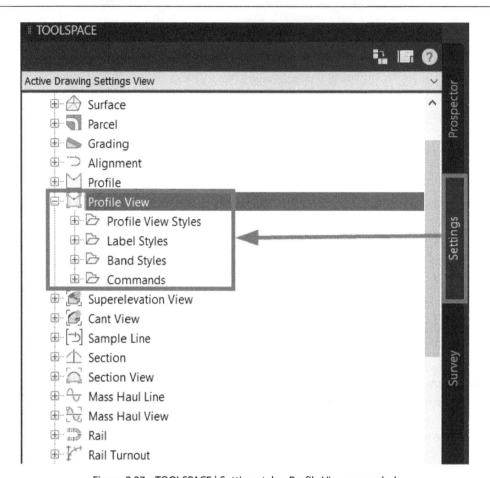

Figure 8.27 – TOOLSPACE | Settings tab – Profile View expanded

Profile View Styles

If we click on the + icon next to **Profile View Styles**, we will then see a comprehensive list of profile label styles that reside in our current drawing.

Once expanded, we'll see the following list of styles that can be applied to profile objects shown in our profile views within our current file for display purposes (also refer to *Figure 8.29*):

- **Station Elevation**: Allows us to create and apply individual station and elevation labels within **Profile View** (refer to *Figure 8.28* for workflow and location)

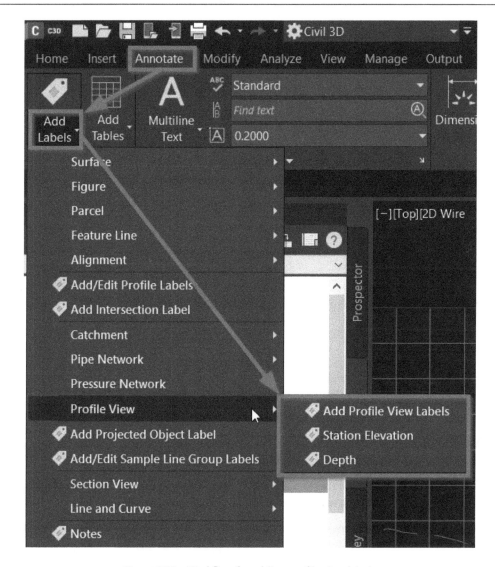

Figure 8.28 – Workflow for adding profile view labels

- **Depth**: Allows us to create and apply individual depth labels based on a two-point identification within our profile view

- **Projection**: Allows us to create and apply individual project labels to objects that have been projected into our profile view

- **Crossing**: Allows us to create and apply individual crossing labels to objects that are displayed and cross our alignment geometry within our profile view

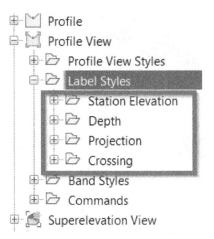

Figure 8.29 – Profile View | Label Styles

Last up are the band styles that we can apply to our profile views. If we click on the + icon next to **Band Styles**, we will then see a comprehensive list of band styles that reside in our current drawing.

Once expanded, we'll see the following list of band styles that can be applied to our profile views within our current file (also refer to *Figure 8.30*):

- **Band Sets**: Allows us to configure and apply groups, or sets, of band labeling to our profile views

- **Profile Data**: Allows us to configure and display station and elevation data for up to two profiles at each station in our profile view banding

- **Vertical Geometry**: Allows us to configure and display geometric data associated with profile tangents and curves displayed in our profile view within our profile view banding

- **Horizontal Geometry**: Allows us to configure and display geometric data associated with alignment tangents, curves, and spirals within our profile view banding

- **Superelevation Data**: Allows us to configure and display transitional data at curves within our profile view banding

- **Sectional Data**: Allows us to configure and display sectional data within our profile view banding

- **Cant Data**: Allows us to configure and display cant data within our profile view banding

- **Speed**: Allows us to configure and display speed data associated with profiles within our profile view banding

- **Pipe Network**: Allows us to configure and display gravity pipe network data associated with our profile view banding

- **Pressure Network**: Allows us to configure and display pressure pipe network data associated with our profile view banding

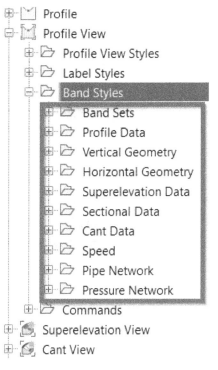

Figure 8.30 – Band Styles

With the overview of all applicable styles that can be applied to our profiles and profile views out of the way, let's jump into getting a bit more familiar with how we can analyze our profile and alignment geometry, all the while getting more comfortable with our design in general.

Further analyzing our profile and alignment geometry

As you may recall, when we were exploring the contextual ribbon associated with alignments back in *Chapter 7, Alignments - The Second Foundational Component to Designs within Civil 3D*, we had a few analysis tools available to us, but we were not able to use them due to there not being a profile present in that particular drawing file.

Now that we've data referenced our alignment objects in the Grading Model.dwg file and have also created profiles and profile views that essentially tie both the horizontal (alignments) and vertical (surfaces) Civil 3D objects together, we're now able to explore these tools.

If we select either our alignment or profile objects, the same analysis tools become available to us in their respective **Contextual** ribbons. That said, let's go ahead and select our **PRF - Subdivision Main Road - Access** profile object to access our Profile **Contextual** ribbon, as shown in *Figure 8.31*.

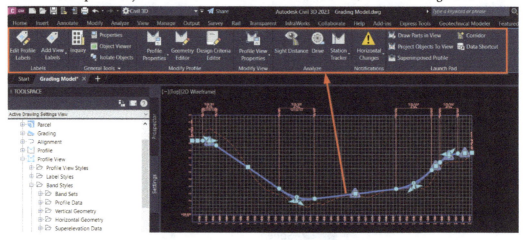

Figure 8.31 – Profile Contextual ribbon

Running from left to right, we have the following panels with tools available to us:

- **Labels**: Grants us access to various annotation tools

- **General Tools**: Grants us access to various inquiry tools that will display information associated with our selected profile

- **Modify Profile**: Grants us access to various tools that allow us to manipulate our profile geometry

- **Modify View**: Grants us access to profile view editing tools

- **Analyze:** Grants us access to various analysis tools to further examine our alignment and profile geometry

- **Notifications**: Grants us the ability to dismiss warnings in our profile views when alignment adjustments are made that affect our profile geometry

- **Launch Pad**: Grants us access to major tools/workflows that will enable further development of our design and collaboration

Let's go ahead and take a closer look at the **Analyze** panel containing various analysis tools. You may have noticed that if we select our alignment, we have the same three tools available to us in the **Analyze** ribbon.

Starting with the **Sight Distance** tool, if we select this option in the **Analyze** panel, a **Sight Distance Check** dialog box will appear. There, we can begin filling out criteria to validate our horizontal and vertical geometry in the sense that we are meeting our project design requirements as a minimum.

We also have the ability to either populate our drawing with obstruction data and/or generate a report that includes calculated findings.

Next up is the **Drive** tool. This is where we can run a vehicle simulation as if we were driving down the centerline of our road alignment and following our design (or existing) profile objects. We have the ability with this tool to change our paths, speed, and the level at which our eyes will be while driving. This tool truly provides a great sense of depth and scale.

The final analysis tool we have available to us is the **Station Tracker** tool. The **Station Tracker** tool allows us to quickly visualize corresponding locations in plan and profile views.

Once the **Station Tracker** tool is initiated, if we hover our mouse pointer inside the profile view, we can see a cross-sectional line appear in the subdivision layout at the same station as our mouse is currently at within the profile view.

Similarly, if we hover our mouse pointer along the alignment, we see a vertical line appear at the same station within the profile view. This tool allows us to quickly identify the best course of correction should there be any issues with our design layouts.

Summary

We have now not only begun our residential subdivision design, but we have also gained some tremendous insights into our three core foundational components that are typically required for any project design within Civil 3D. In this chapter specifically, we learned how profiles tie the first two foundational components, surfaces and alignments, together. We've also begun to understand the value each brings to the table and how all three foundational components are dynamically linked together, allowing us to focus on design-authoring workflows rather than manually updating these particular objects in multiple locations.

In the next section, we'll continue progressing with our residential subdivision design, all while discovering the many additional tools available to us within Civil 3D that will help us along our journey. We are still in the very early stages of designing our residential subdivision design, so you can only imagine how much more there is to unpack.

One thing to keep in mind throughout our journey, though, is that just like our surface, alignment, and profile objects have the ability to maintain a dynamic link, we will be able to continue applying similar dynamic links to our additional design objects along the way.

Maintaining this dynamic linking design workflow will minimize rework, keeping us as efficient as possible and staying on track with our project progression and schedule.

Part 3:
Leveraging Design-Specific Tool Belts

In this next part, we will dive deeper into all of the design- and modeling-based tools and functionality built into Civil 3D. Although all of the tools and functionality are multipurpose and cross-functional, we have prepared the learning content to be categorized into three major types of design applications typically seen across the AEC industry.

The following chapters are included in this section:

- *Chapter 9, Land Development Tool Belt for Everyday Use*
- *Chapter 10, Roadway Modeling Tool Belt for Everyday Use*
- *Chapter 11, Advanced Roadway Modeling Tool Belt for Everyday Use*
- *Chapter 12, Utility Modeling Tool Belt for Everyday Use*

9

Land Development Tool Belt for Everyday Use

In *Part 1, Getting Acquainted with Civil 3D and Starting Your Next Project Up for Success*, we gained some insight and understanding of how Civil 3D operates as we discussed many of the foundational settings, configurations, and workflows necessary to begin a design project within the Civil 3D environment.

In *Part 2, Designing and Modeling with Civil 3D from Start to Finish*, we learned how properly setting up a survey model from the get-go is critical to running a smooth design. We've also gained some insight into how surfaces, alignments, and profiles are all tied to each other dynamically and are the core ideas required to progress into our residential subdivision design moving forward.

In this chapter, we'll continue to progress through our residential subdivision design while also gaining additional insight and understanding of which other tools are available inside of Civil 3D and how we can leverage all while maintaining a dynamic link between these objects.

We'll also begin to see firsthand how maintaining dynamic linking between all modeled objects in our design will save us many steps and unnecessary time later on down the road in the event that we need to pivot and make some adjustments in our design due to various unforeseen circumstances.

Furthermore, we'll begin unwrapping our major land development tools that are available to us within Civil 3D. These tools will allow us to continue building on top of our foundational objects and take the next step in modeling a residential subdivision design layout. That said, in this chapter, we'll be covering the following topics:

- Creating and managing parcels
- Creating and managing sites
- Leveraging grading tools for our site design

With that, let's go ahead and open up Civil 3D, or go to your start screen if already open, and create a new drawing using similar steps outlined in *Chapter 7, Alignments - The Second Foundational Component to Designs within Civil 3D*. We can use our `Company Template File.dwt` file located in `Practical Autodesk Civil 3D 2024\Chapter 9` and select **Open** in the lower right-hand corner of the **Select Template** dialog box. Once our new file is created, we'll want to save it as our `Site Plan Reference.dwg` file to our `Practical Autodesk Civil 3D 2024\Chapter 9\Reference` location.

As discussed back in *Chapter 3, Sharing Data within Civil 3D*, reference files are intended to contain/ represent 2D geometry and static elements and annotation. Reference files would include content such as surveyed planimetrics, civil site plan geometry, erosion control BMPs, and so on.

Although parcels are considered Civil 3D objects, they are not able to be data referenced across multiple files. When we get to the sections covering sites and grading tools, we'll want to jump back into our `Grading Model.dwg` file to continue progressing the modeled design objects that will require further coordination via data referencing.

With that, let's go ahead and attach our `Survey Model.dwg` file as an overlay, contained within our `Practical Autodesk Civil 3D 2024\Chapter 9\Model` location. Then, we'll want to jump back into our **Prospector** tab in our Toolspace, and then set the working folder of our data shortcuts project to the `Practical Autodesk Civil 3D 2024\Chapter 9` location and select the `C3D_2024_123456_Data_Shortcuts` project, as shown in *Figure 9.1*:

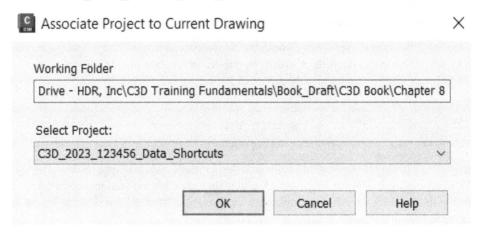

Figure 9.1 – Associate Project to Current Drawing dialog

After our Civil 3D data shortcuts project has been associated with our current file, we can then safely create data references of **ALG - Subdivision Main Road - Access** and **ALG - Subdivision Side Road - Cul-De-Sac** alignments into our current `Site Plan Reference.dwg` file.

To do this, we need to expand our **Alignments** category and our `Proposed Conditions` folder, right-click on each object (shown in *Figure 9.2*) and select the **Create Reference** option for each:

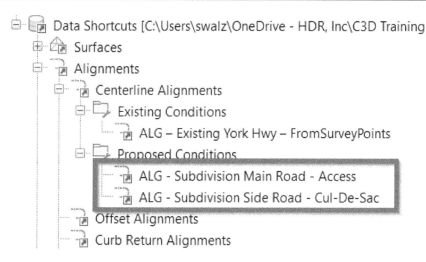

Figure 9.2 – Creating data references of identified Civil 3D objects

Then, with our `Site Plan Reference.dwg` file set up for us to include all objects we have available to us to represent and reference the existing environment, along with a few objects of the future built environment, we are now ready to continue designing our residential subdivision layout.

Technical requirements

It's important to note that Autodesk's Civil 3D can oftentimes be very taxing on your computer. There is a lot of processing that goes on with modeled design elements, even in the background, that enables the dynamic (connected) capabilities to occur throughout the **Building Information Modeling** (BIM) design lifecycle.

In turn, many technical requirements need to be considered to allow Autodesk's Civil 3D to operate at its full potential. We'll review the minimum requirements that Autodesk recommends, with a few of my suggestions added to increase efficiency and speed throughout the BIM design process:

- **Operating system**: 64-bit Microsoft Windows 10

- **Processor**: 3+ GHz

- **Memory**: 16 GB RAM (I suggest going with either 64 GB or 128 GB)

- **Graphics card**: 4 GB (I suggest going with 8+ GB)

- **Display resolution**: 1980 x 1080 with True Color

- **Disk space**: 16 GB

- **Pointing device**: MS Mouse compliant

The exercise files for this chapter are available at `https://packt.link/UoiPn`

Creating and managing parcels

In this section, we'll begin to explore and understand how to utilize the parcel tools available to us within Civil 3D to support our residential subdivision design. We'll begin by creating our overall parcel object that will combine the two existing parcels from our survey.

Using the `NCOPY` command (type this at the command line), we'll go ahead and copy our linework showing only the outermost parcel boundary from our `Survey Model.dwg` external reference into our `Site Plan Reference.dwg` file, as shown in *Figure 9.3*:

Figure 9.3 – NCOPY outermost parcel linework

Once this linework has been extracted from our `Survey Model.dwg` file and placed into our `Site Plan Reference.dwg` file, we'll then want to isolate these objects and use the `JOIN` command (type this at the command line) to create one continuous polyline of the outermost parcel linework.

The next step we'll want to take is to convert this linework to a parcel object. To access our parcel creation tools, we'll want to activate our **Home** ribbon along the top of our Civil 3D session, go to our **Create Design** panel, and click on the down arrow next to where it says **Parcel** (refer to *Figure 9.4*):

Figure 9.4 – Parcel tools

Let's go ahead and select the **Create Parcel from Objects** tool and select the new linework we just extracted from our `Survey Model.dwg` file. Once selected, hit the *Enter* key and we'll then be prompted with a **Create Parcels - From objects** dialog box, as shown in *Figure 9.5*:

Figure 9.5 – Create Parcels From objects dialog box

In our **Site** field, we'll want to go ahead and create a new site by selecting the down arrow to the right of the icon next to the field and selecting the **Create New** option (refer to *Figure 9.6*):

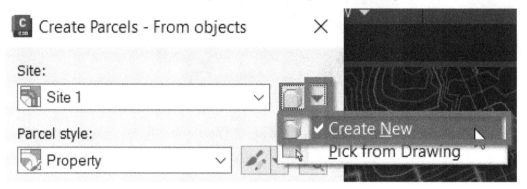

Figure 9.6 – Creating a new site

Once selected, we'll fill out the following fields, also shown in *Figure 9.7*, and then hit the **OK** button in the lower right-hand corner of the **New Site Creation** dialog box:

- **Name: SIT - Proposed – Overall**

- **Description: Overall Residential Subdivison Boundary**

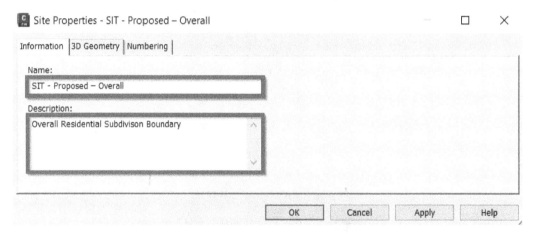

Figure 9.7 – New Site Creation dialog box

We'll then make the following selection in our **Create Parcels From objects** dialog box, also shown in *Figure 9.8*, and then click on the **OK** button:

- **Site: SIT - Proposed – Overall**

- **Parcel style: Property**

- **Area label style**: **Name Area & Perimeter**
- **Line segment label style**: **Bearing over Distance**
- **Curve segment label style**: **Delta over Length and Radius**
- **Automatically add segment labels**: *Check*
- **Erase existing entities**: *Check*

Figure 9.8 – Create Parcels From Objects dialog box

Next, we'll want to create a right of way within our newly created parcel that follows our data-referenced **ALG - Subdivision Main Road - Access** and **ALG - Subdivision Side Road - Cul-De-Sac** alignments. Before creating our right of way, though, we'll need to make a slight adjustment and move our data-referenced alignments into our newly created **SIT - Proposed – Overall** site.

To do so, we'll select both data-referenced **ALG - Subdivision Main Road - Access** and **ALG - Subdivision Side Road - Cul-De-Sac** alignments, right-click on our mouse, and select the **Move to Site…** option, as shown in *Figure 9.9*:

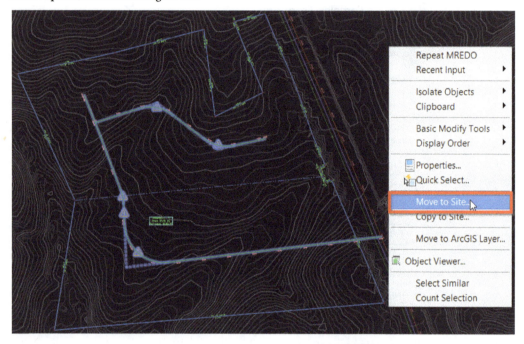

Figure 9.9 – Moving alignments to the site

When our **Move to Site** dialog box appears, we'll select the **Destination site** type as **SIT - Proposed – Overall**, check the **Alignments** boxes, and then click on the **OK** button, as shown in *Figure 9.10*:

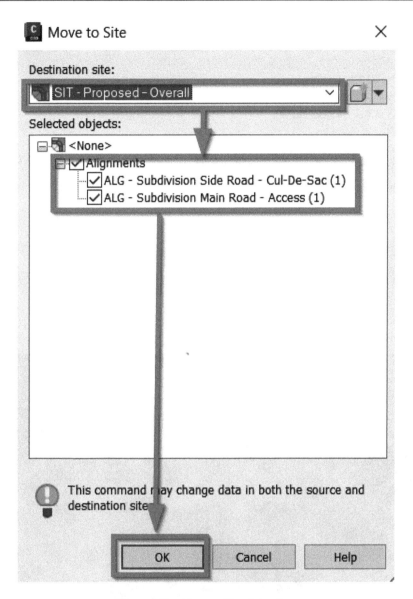

Figure 9.10 – Move to Site dialog box

Now, going back up to our parcel creation tools in our **Home** ribbon, we'll select the **Create Right of Way** tool. When prompted to select our parcel, we'll need to select our **Parcel** label in the center of our newly created parcel, instead of the outermost boundary linework, and hit *Enter* to accept our selection.

After hitting the *Enter* button on our keyboard, a **Create Right of Way** dialog box will appear, where we'll accept the default values and selections, as shown in *Figure 9.11*:

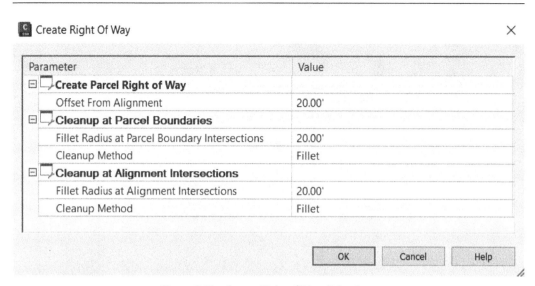

Figure 9.11 – Create Right of Way dialog box

After clicking on the **OK** button in our **Create Right of Way** dialog box, our proposed parcel layout should look similar to that shown in *Figure 9.12*:

Figure 9.12 – Site plan reference model after right-of-way creation

With our outer parcel boundary and our right-of-way parcel created, we can now use the parcel creation tools to lay out our individual lots within our residential subdivision design. Going back up to our **Home** ribbon, then to the **Parcel** dropdown, let's now select the **Parcel Creation Tools** option, at which point our **Parcel Layout Tools** toolbar will appear:

Figure 9.13 – Parcel Layout Tools toolbar

Running through the tools as numbered in *Figure 9.13*, we have the following:

1. **Create Parcel**: Allows us to create parcels based on criteria defined within *item 13*, **Expand the Toolbar**

2. **Add Fixed Line – Two Points**: Allows us to manually create a fixed parcel line based on two points

3. **Add Fixed Curve**: Allows us to manually create a fixed curve based on either two or three points

4. **Draw Tangent – Tangent with no Curves**: Allows us to manually create tangents

5. **Line Tools**: Allows us to either create or edit parcel lines

6. **Point of Intersection Tools**: Allows us to add, remove, and/or break apart **points of intersection (PIs)**

7. **Delete Sub-Entity Tools**: Allows us to delete sub-entities based on manual selection

8. **Parcel Union Tools**: Allows us to join multiple parcels

9. **Pick Sub-Entity**: Allows us to select a parcel sub-entity and display its parameters

10. **Sub-Entity Editor**: Allows us to display parameters of identified sub-entities

11. **Undo**: Allows us to undo the previous command(s)

12. **Redo**: Allows us to reapply or redo commands that were undone

13. **Expand the Toolbar**: Allows us to define criteria for *item 1*, **Create Parcel**

Now that we have a high-level understanding of which parcel creation tools we have available to us, let's start putting some of them to practice. Before getting started, though, we'll need to look up our local town/city/county/state zoning codes to be able to fill in our parcel criteria.

An example of the parcel criteria we'll be using for this particular exercise can be found at `https://www.yorkcounty.gov/DocumentCenter/View/82/Zoning-Ordinance-Summary-Table-PDF?bidId=`. Once your project's location is determined, we can then go ahead and use the **Expand the Toolbar** option within our **Parcel Layout Tools** toolbar to define our criteria for parcel generation, also shown in *Figure 9.14*:

- **Minimum Area**: `33000.00 Sq. Ft.`
- **Minimum Frontage**: `50.00'`
- **Use Minimum Frontage At Offset**: `Yes`
- **Frontage Offset**: `15.00'`
- **Minimum Width**: `130.00'`
- **Minimum Depth**: `30.00'`
- **Use Maximum Depth**: `No`
- **Maximum Depth**: `500.00'`
- **Multiple Solution Preference**: `Use shortest frontage`
- **Automatic Mode**: `On`
- **Remainder Distribution**: `Redistribute remainder`

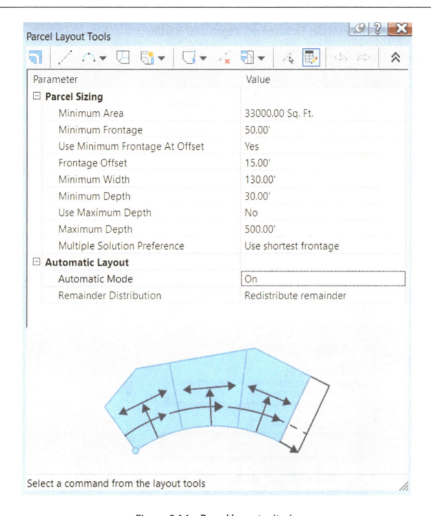

Figure 9.14 – Parcel layout criteria

After filling out our parcel criteria, we'll then select the down arrow next to our **Line** tool icon (listed as number **5** in *Figure 9.13*) and select the **Slide Line Create** tool to begin our automatic parcel generation throughout our residential subdivision site. Once the **Slide Line Create** tool has been selected, a **Create Parcels Layout** dialog box will appear, at which point we'll want to make the following selections (also shown in *Figure 9.15*):

- **Site**: **SIT - Proposed – Overall**
- **Parcel style**: **Single-Family**
- **Area label style**: **Parcel Name**
- **Line segment label style**: **Bearing over Distance**

- **Curve segment label style**: Delta over Length and Radius
- **Automatically add segment labels**: *Check*

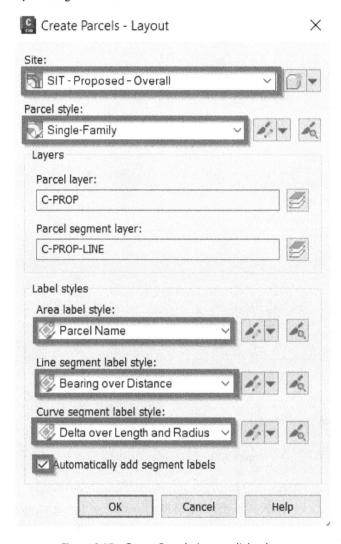

Figure 9.15 – Create Parcels Layout dialog box

After our fields have been filled and we select the **OK** button within the **Create Parcels Layout** dialog box, we'll want to bring our attention down to the command line and follow the steps listed out here:

1. First, we'll be prompted to select the starting point on the frontage, which refers to our street right-of-way frontage. Let's go ahead and snap to the endpoint of our right-of-way parcel, on the north side, closest to York Highway, as shown in *Figure 9.16*:

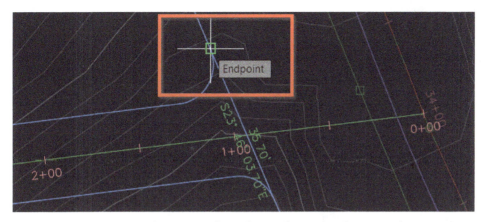

Figure 9.16 – Snapping to the endpoint of our right-of-way parcel

2. After snapping to the northern endpoint of our right-of-way parcel shown in *Figure 9.16*, we'll then be prompted at our command line to select the endpoint of our frontage, which will be our southern endpoint.

Before snapping directly to the southern endpoint, we'll need to drag our mouse from right to left along the entire length of our right-of-way parcel to ensure that we are defining the full path, which will be recognized as street frontage in our automatic parcel creation process, as shown in *Figure 9.17*:

Figure 9.17 – Defining path of street frontage

3. Once our path has been highlighted as shown in *Figure 9.17*, we'll then want to snap to the endpoint of our right-of-way parcel, this time on the south side, closest to York Highway, as shown in *Figure 9.18*:

Figure 9.18 – Snapping to the endpoint of our right-of-way parcel

4. After snapping to the endpoint on the southern side, we'll then be prompted to specify the angle in our command line. For this prompt, we'll just hit the *Enter* key on our keyboard to bypass and accept the default.

5. After hitting the *Enter* key, Civil 3D will run through the process of automatically laying out our **Single-Family** parcels throughout our residential subdivision design, as shown in *Figure 9.19*:

Figure 9.19 – Automatic parcel layout

6.   Finally, we'll then be asked in our command line if we'd like to accept the results, at which point we'll type YES and hit the *Enter* key to finish the automatic parcel layout process, with the final result appearing similar to that shown in *Figure 9.20*:

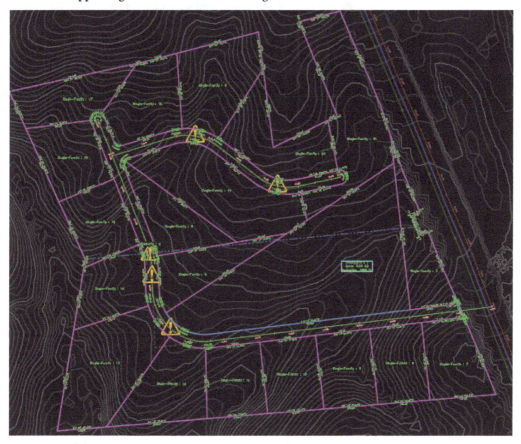

Figure 9.20 – Automatic parcel layout results

Congratulations! We have now created our **Single-Family** parcels, leveraging automated workflows within Civil 3D, to lay out our residential subdivision design. If adjustments need to be made to our parcel layout thereafter, we'd want to pull up our **Parcel Layout Tools** toolbar again, go back to our **Line Tools** dropdown (identified as number **5** in *Figure 9.13*), and select the **Single Line – Edit** tool to adjust our parcel lines accordingly.

Creating and managing sites

As we were creating parcels in the previous section, you may have noticed that during the parcel creation process we had to associate our parcels with a site, and we ended up creating a site called **SIT - Proposed – Overall**. In this section, we'll dive a bit deeper to explore and understand how to continue leveraging and incorporating sites into our residential subdivision design.

With that, let's go ahead and open up our `Grading Model.dwg` file located within our `Practical Autodesk Civil 3D 2024\Chapter 9\Model` location. Once opened, you'll notice that I've already prepped the file to contain polylines representing our **Single-Family** parcel boundary lines created in the previous section, along with building setback lines offset `20'` from each lot boundary.

Using the following basic AutoCAD commands, I was able to convert our **Single-Family** parcel boundaries to polylines and carry them into our `Grading Model.dwg` file:

1. Begin by saving the `Site Plan Reference.dwg` file.
2. In our `Site Plan Reference.dwg` file, we'll isolate parcels (including **Single-Family** and **Right-of-Way** parcels).
3. **Window** select all parcels after isolation.
4. Type `explode` in the command line to explode parcels.
5. Type `flatten` into the command line to flatten parcels (to ensure that all linework is at elevation 0).
6. Leverage the `BPOLY` command (type this at the command line) to create polyline boundaries for each individual parcel.
7. Select the new polylines that we just created.
8. Right-click on the mouse and select the **Clipboard | Copy with Base Point** option.
9. Type in `0,0,0` as the base point coordinates at the command line.
10. Jump into our `Grading Model.dwg` file, right-click, and select the **Clipboard | Paste** option.
11. Type in `0,0,0` as the base point coordinates at the command line.
12. Offset boundary lines internal to each lot by 20 feet, except for our Open Space lot (shown hatched in *Figure 9.21*).
13. The final appearance should look similar to that shown in *Figure 9.21*:

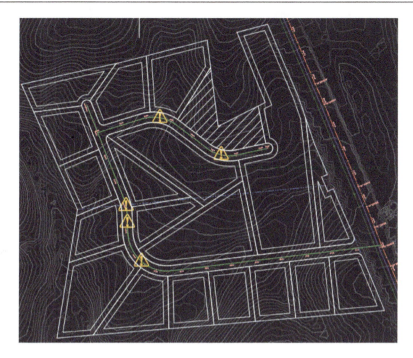

Figure 9.21 – Grading Model.dwg file after polylines representing parcels are copied and pasted

Bringing our attention back to the Toolspace, we'll want to select the **Prospector** tab, navigate to our sites (as shown in *Figure 9.22*), right-click on **Sites**, and select **New**:

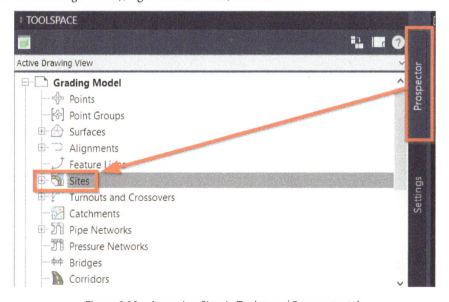

Figure 9.22 – Accessing Sites in Toolspace | Prospector tab

After selecting **New**, a **Site Properties** dialog box will appear, at which point we'll fill out the following fields accordingly (also shown in *Figure 9.23*), and then click the **OK** button in the lower right-hand corner of the dialog box:

- **Name**: **SIT - Proposed – Building Pads**

- **Description**: **Buildings Pad Grading**

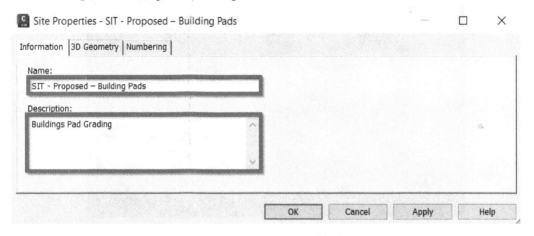

Figure 9.23 – Site Properties dialog box

It's important to note that sites by themselves are managed directly in the drawing file they are created within, with a few exceptions. If we go back to our Toolspace, select the **Prospector** tab, and expand our **Sites** category and newly created **SIT - Proposed – Building Pads** site, we can see we have a few types of Civil 3D objects that are able to be associated with each individual site created.

As shown in *Figure 9.24*, we have the following Civil 3D objects able to be associated with sites: **Alignments**, **Feature Lines**, **Grading Groups**, and **Parcels**:

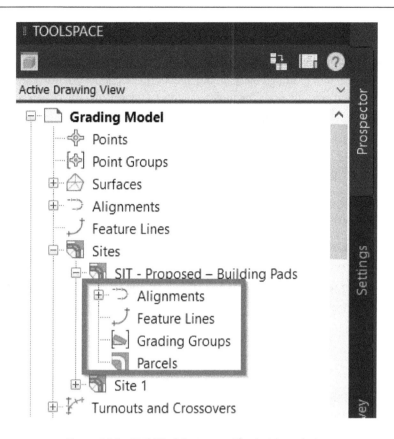

Figure 9.24 – Civil 3D objects associated with each site

> **Note**
>
> Alignments are the only object in Civil 3D 2024 that can be data referenced into other files from within our sites. When data referencing alignments that are associated with sites into other drawing files, they will automatically create the site that they are associated with, and the alignments will be contained within.

Sites alone do not provide a ton of function, but Civil 3D objects contained within add tremendous value to our designs. That said, sites add another level of Civil 3D object management that allows us to reference objects from other sites as needed.

As an example, if we were to create feature lines in one site, we would be able to set our grading groups that utilize those feature lines to target or reference components in other sites. This concept will become a little clearer as we progress throughout the next sections and chapters.

Leveraging grading tools for our site design

Now that our new **SIT - Proposed – Building Pads** site has been created, we can now go ahead and start adding feature lines and grading objects to our site. In this section, we'll learn how to leverage multiple feature lines and grading modeling tools to support a typical civil site design and analysis to further our residential subdivision design.

For starters, we'll begin by converting our setback lines to feature lines that will ultimately be associated with our **SIT - Proposed – Building Pads** site. To convert our polylines representing the setback lines, we'll go up to our **Home** ribbon and locate our feature line tools located within our **Create Design** panel.

Once located, we'll click on the down arrow next to **Feature Line** to pull down the following list of feature line tools available to us (also shown in *Figure 9.25*):

- **Create Feature Line**: Allows us to manually create new feature lines by defining PIs and **points of vertical intersection (PVIs)**

- **Create Feature Lines from Objects**: Allows us to convert existing lines, arcs, polylines, and 3D polylines to feature lines

- **Create Feature Lines from Alignment**: Allows us to convert existing alignments to feature lines

- **Create Feature Line from Corridor**: Allows us to convert feature lines from existing corridor-modeled breaklines

- **Create Feature Line from Stepped Offset**: Allows us to create a feature line based on a defined horizontal and vertical offset from an identified feature line, survey figure, polyline, and/or 3D polyline

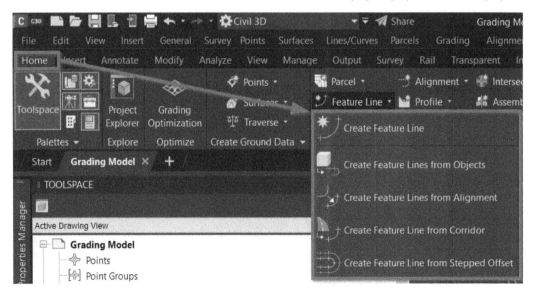

Figure 9.25 – Feature line tools

To actually convert our setback polylines to feature lines, we'll want to select **Create Feature Line** from the **Objects** tool. Then, once the command has been initiated, we'll go ahead and select all of our setback polylines, as shown in *Figure 9.26*:

Figure 9.26 – Creating feature lines from an object: selection of setback polylines

After all polylines have been selected, we'll simply hit the *Enter* key on our keyboard to accept our selection set, where we'll then have a **Create Feature Lines** dialog box appear.

Let's go ahead and fill out the following fields and make the following selections, as shown in *Figure 9.27*:

- **Site: SIT - Proposed – Building Pads**
- **Style**: Check the box and select the **Platform Edge** style
- **Erase existing entities**: *Check*
- **Assign elevations**: *Check*

Figure 9.27 – Create Feature Lines dialog box

Since we had checked the box next to **Assign elevations**, we'll then be prompted with an **Assign Elevations** dialog box after clicking the **OK** button in the lower portion of the **Create Feature Lines** dialog box.

In the **Assign Elevations** dialog box, we'll want to select the option to assign elevations from our **SRF Existing Grade – FromSurveyPoints** surface model, and also check the box next to **Insert intermediate grade break points**, and then click the **OK** button in the lower portion of the **Assign Elevations** dialog box, shown in *Figure 9.28*:

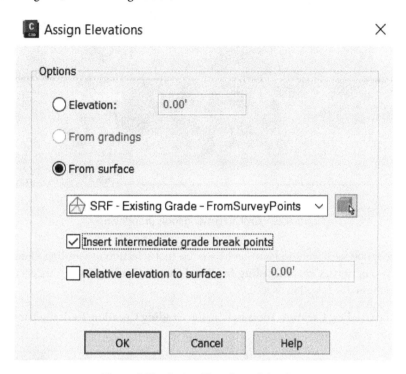

Figure 9.28 – Assign Elevations dialog box

> **Note**
>
> By checking the box next to **Insert intermediate grade break points**, we are adding PVIs along the perimeter of each of the polylines where they cross the **triangulated irregular network (TIN)** lines of our **SRF - Existing Grade – FromSurveyPoints** surface model. This will allow us to generate a very tight and accurate design surface model as we progress our residential subdivision layout.

With our feature lines created, we'll now want to explore the grading tools that we have available to us within Civil 3D. Going back up to our **Home** ribbon and **Create Design** panel, let's click the down arrow next to **Grading** to access our grading tools, as shown in *Figure 9.29*:

Figure 9.29 – Civil 3D grading tools dropdown

The key grading option we'll want to focus on here is the first selection of **Grading Creation Tools**. The additional two options, **Create Grading Infill** and **Create Grading Group** are accessible with the **Grading Creation Tools** option as well.

After selecting the **Grading Creation Tools** option, the **Grading Creation Tools** toolbar will appear, as shown in *Figure 9.30*:

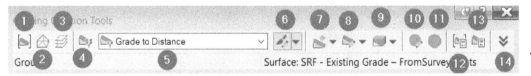

Figure 9.30 – Grading Creation Tools toolbar

Running through the tools as numbered in *Figure 9.30*, we have the following:

1. **Set the Grading Group**: Allows us to select and/or create a grading group that our grading objects will be associated with

2. **Set the Target Surface**: Allows us to select a surface in our current file that we will target for daylighting purposes

3. **Set the Grading Layer**: Allows us to manually set the grading layer on which our grading objects will be placed

4. **Select a Criteria Set**: Allows us to utilize a predefined grading criteria set

5. **Select a Grading Criteria**: Allows us to specify which type of grading object we intend to create next

6. **Grading Criteria Manager**: Allows us to create and modify grading criteria

7. **Grading Type Selection**: Allows us to change the grading type from a predefined **Grading Criteria** selection to an **Infill** or **Transition between two grading objects and slopes**

8. **Grading Editing Tools**: Allows us to modify grading objects

9. **Grading Volume Tools**: Allows us to access grading volume tools

10. **Grading Editor**: Allows us to edit grading objects

11. **Elevation Editor**: Allows us to edit elevations of feature lines

12. **Grading Group Properties**: Allows us to edit grading group properties

13. **Grading Properties**: Allows us to edit properties associated with grading objects

14. **Expand the Toolbar**: Allows us to define criteria as relates to grading methods, slope projection, and conflict resolution

Moving forward, we'll want to create grading objects for each parcel so that we can establish a flat area inside of all of the setback lines. This flat area will essentially represent the building pads upon which our houses will be built. That said, we'll want to take note of the approximate elevation on the street side within each of our setback feature lines and create grading objects to target those elevations.

To identify the approximate elevation, we'll simply hover our mouse over a decent location for a second or so, at which point Civil 3D object references will list out next to our cursor, one of which will reference our **SRF - Existing Grade – FromSurveyPoints** surface model, along with the elevation that our cursor is hovering over (refer to *Figure 9.31* for an example of what this may look like):

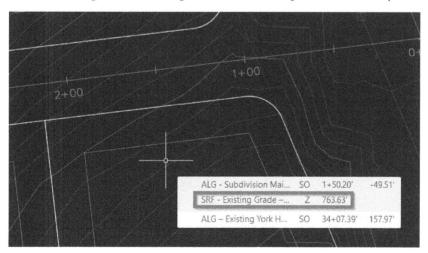

Figure 9.31 – Identifying the approximate elevation on the roadside of the setbacks

Using the following workflow and detailed steps for our southeastern parcel, we'll begin applying grading objects and generating a proposed surface model using these setback feature lines:

1. Let's go ahead and change our grading criteria selection from **Grade to Distance** to **Grade to Elevation** by using the down arrow dropdown in our **Select Grading Criteria** box (listed as number **5** in *Figure 9.30*).

2. Once changed to **Grade to Elevation**, we'll then select our **Grading Type Selection** icon (listed as number **7** in *Figure 9.30*), to initiate the creation process.

3. You will then be prompted to select a site that we want to associate our grading objects with. When the dialog box appears, we'll want to select our **SIT - Proposed – Building Pads** site and select the **OK** button.

4. Once the **OK** button has been selected, a **Create Grading Group** dialog box should appear. Let's go ahead and fill out the following fields and make the following selections (also shown in *Figure 9.32*), and then click the **OK** button toward the bottom of the dialog box:

   - **Name**: SGG - Proposed – Building Pads

   - **Description**: Proposed Building Pads Grading Objects

   - **Automatic surface creation**: *Check*

   - **Surface style**: Contours 1' and 5' (Design)

   - **Tessellation spacing**: 10.00'

   - **Tessellation angle**: 3.0000 (d)

   - **Volume base surface**: Check and select **SRF - Existing Grade – FromSurveyPoints**

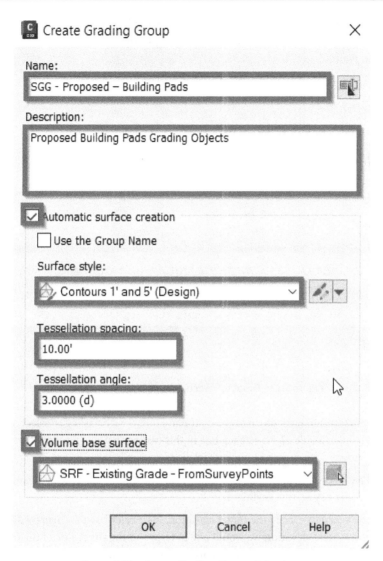

Figure 9.32 – Create Grading Group dialog box

5. After selecting the **OK** button, we'll be prompted with one more dialog box, this time being a **Create Surface** dialog box. In the **Create Surface** dialog box, we'll want to give our new surface a name in close alignment with the grading group that it's referencing. That said, let's go ahead and call it **SGG - Proposed Grade – Building Pads** and then select the **OK** button.

6. If we bring our attention down to our command line, we'll notice that we are being prompted to select a feature line that we would like to use as a starting point to build our grading objects. Starting with our first lot in the southeastern portion of our residential subdivision, we'll select our **Setback** feature line.

7. We'll then see a new message appear on our screen that indicates that the feature line should be weeded. Go ahead and select the **Continue Grading Without Feature Line Weeding** option, as shown in *Figure 9.33*:

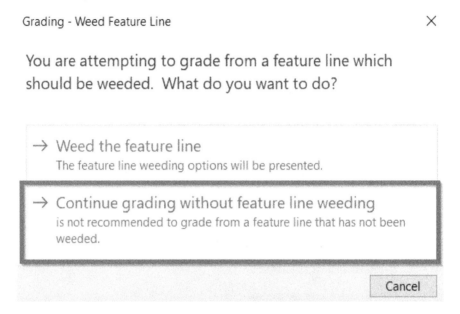

Figure 9.33 – Grading - Weed Feature Line message

8. Bringing our attention back down to the command line, we'll be asked to select a grading site. We'll want to click inside of the setback line in this case so that the grading object created will generate a surface within each of our setback lines.

9. Next, we'll be asked if we want to apply to the entire length. In our case, we want to type YES and hit the *Enter* key on our keyboard.

10. We'll then be asked to specify the target elevation. If we refer back to *Figure 9.31*, we'll notice that the approximate elevation we would like to target on the street side of the parcel is listed as 763.63. Let's type this value at the command line and hit *Enter*.

11. Next, we'll be asked to specify a **Cut Format** type, where we can accept the default of **Slope**.

12. Then, we'll be asked to specify a **Cut Slope** value, where we'll type in 2:1 (rise over run) and hit *Enter*.

13. The same will then be asked for **Fill** scenarios. Let's make the same selection of **Slope**, and then type 2:1 as the value and hit *Enter*.

After all steps have been completed, we'll notice that our grading objects and surface have been created within the setback feature lines of our southeastern parcel, as shown in *Figure 9.34*:

Figure 9.34 – Final grading object and grading surface for our southeastern parcel

Let's use these same steps applied to our first parcel to continue our grading object and grading surface creation throughout our site, with the exception of our northeastern lot since that appears to be our largest lot with the biggest elevation difference throughout.

For this particular lot, we'll want to target an elevation relatively close to the existing elevation in the middle of our lot, closest to the cul-de-sac location, which is approximately 744.00. Let's use this target elevation for this lot in particular.

After applying these steps to all setback feature lines, the last thing we'll need to do is apply an infill grading object inside each of the grading objects displayed in our lots. Going back up to our grading tools, we'll select the down arrow in our **Grading Type Selection** icon (listed as number **7** in *Figure 9.30*) and select the **Infill** option. After being selected, we'll proceed with clicking our left mouse button inside each of the grading objects previously created. As we click inside these areas, you will notice that the boundaries will be highlighted, and a green diamond will appear at the centroid.

After applying these steps to all setback feature lines, the final product should look similar to that displayed in *Figure 9.35*:

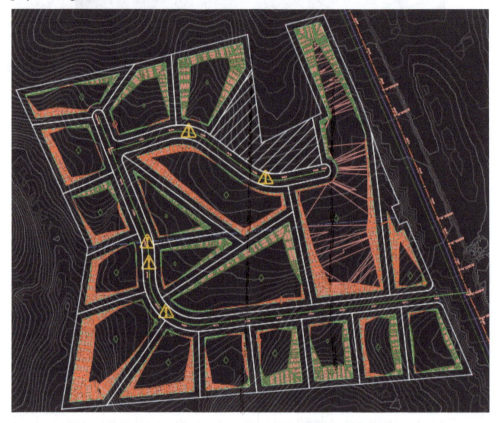

Figure 9.35 – Final grading object and grading surface for our residential subdivision

Summary

We have taken some significant steps throughout this chapter toward developing our general arrangement, and associated modeled objects, to progress our residential subdivision design. We've learned how to create right of ways along with individual parcels, and even incorporated some level of automation while doing so. We then jumped into sites and grading tools to establish building pad locations for us to use when it comes time to locate individual houses throughout our residential subdivision design.

In our next session, we'll replace our land development toolbelt with our roadway modeling toolbelt to continue progressing our residential subdivision. In addition to understanding how we can utilize roadway modeling tools, we'll also be able to tie our roadway surface models into our land development surface models that we have developed, resulting in a federated proposed residential subdivision surface model that will be used for final display purposes.

10
Roadway Modeling Tool Belt for Everyday Use

In the previous chapter, we had an opportunity to unwrap all of the Land Development-focused tools available to us, and also learned some practical applications of how we can utilize the **Parcel**, **Site**, and **Grading** tools to further progress our Residential Subdivision design. Not only did we learn how to use these particular tools to create design objects within Civil 3D but we also began to understand how best to manage these objects and the significance each one provides to the overall design.

In this chapter, we'll continue to progress through our Residential Subdivision design while also gaining additional insight into and understanding of the Roadway Modeling tools that we have available to us inside of Civil 3D and how we can utilize them while maintaining a dynamic link between all modeled objects.

That said, as we begin exploring our major Roadway Modeling tools, we will continue building on top of our Foundational and Land Development model objects as we take the next steps in modeling our Residential Subdivision design layout.

In this chapter, we'll be covering the following topics:

- Creating and managing assemblies
- Creating and modifying corridors
- Creating and modifying intersections and cul-de-sacs
- Creating surfaces from corridors

With that, let's go ahead and open up our `Grading Model.dwg` file located within our `Practical Autodesk Civil 3D 2024\Chapter 10\Model` location. Once opened, you'll notice that we'll be starting this chapter pretty much where we left off in the previous chapter, with the display of our model looking similar to that shown in *Figure 10.1*:

Figure 10.1 – Final grading object and grading surface for our Residential Subdivision

Technical requirements

The exercise files for this chapter are available at `https://packt.link/UoiPn`

Creating and managing assemblies

In this section, we'll begin to learn all about assemblies and why they are considered to be the building blocks of all transportation-focused designs. Although a Residential Subdivision wouldn't necessarily be considered a transportation design, our project requires us to design two roads, two intersections, one dead end, and one cul-de-sac. That said, assemblies in our design scenario are still going to be foundational modeled objects that will be utilized to execute an accurate roadway design.

Without further ado, let's jump into our **Assembly Creation and Management** tools. We can access our basic **Assembly** tools by going back up to the **Home** ribbon, navigating to the **Create Design** panel, and selecting the down arrow next to **Assembly**.

As shown in *Figure 10.2*, we can see that we have two assembly tools available to us: **Create Assembly** and **Add Assembly Offset**:

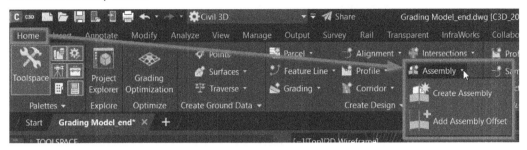

Figure 10.2 – Home | Create Design | Assembly

The **Create Assembly** tool allows us to do just that… create an assembly. The **Add Assembly Offset** tool is most commonly used when a service road needs to be designed that is offset from our main centerline alignment.

In our case, since we are only focusing on main roadways throughout our Residential Subdivision design layout, we will want to make the **Create Assembly** selection to start creating the assembly.

Once selected, we'll be presented with the **Create Assembly** dialog box, at which point we'll fill in the fields available as follows (also displayed in *Figure 10.3*), and then click on the **OK** button in the lower portion of the dialog box:

- **Name: ASM – Subdivision Main Road – Access**
- **Description: Subdivision access road from York Hwy**
- **Assembly Type: Undivided Crowned Road**
- **Assembly style: Basic**
- **Code set style: All Codes**

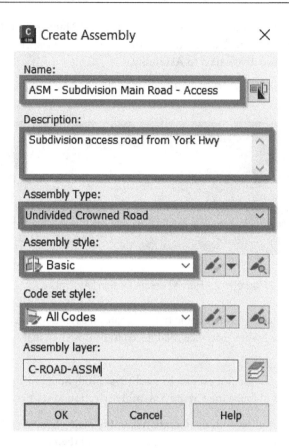

Figure 10.3 – The Create Assembly dialog box

After selecting the **OK** button, we must go back to the command line, where we'll be asked to specify an **Assembly Baseline Location**. What I tend to do in these cases is specify my assembly baseline location near the profile views so that I can group horizontal and vertical components relatively close to each other.

That said, choose a location that you prefer (above, below, besides, and so on) next to your current profile views to place your assembly baseline. Once a location has been identified, Civil 3D will automatically zoom into that location so that only the assembly baseline is visible in your view, as shown in *Figure 10.4*:

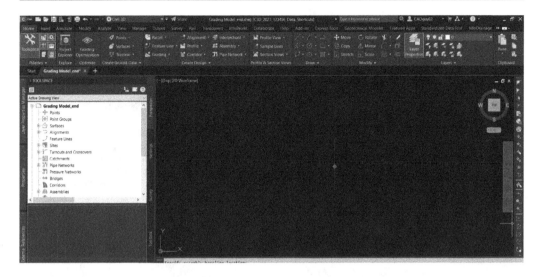

Figure 10.4 – Assembly baseline placed in the model space

Now, if we select our newly placed assembly baseline with our mouse, we'll notice that the **Assembly Contextual** ribbon will appear, as shown in *Figure 10.5*. Running from left to right, we have the following panels with tools available to us:

- **General Tools**: This grants us access to various inquiry and display tools that will surface information and manipulate the appearance of our assembly objects

- **Modify Subassembly**: This grants us access to tools that will allow us to further manipulate the subassemblies that are attached to that particular assembly

- **Modify Assembly**: This grants us access to tools that will allow us to further manipulate our assembly objects

- **Launch Pad**: This grants us access to tools that will further progress and build out our assemblies and develop a corridor model within our current file:

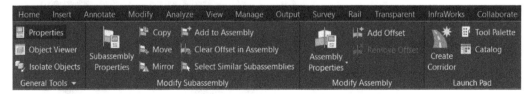

Figure 10.5 – The Assembly Contextual ribbon

Going into the **Launch Pad** panel, let's go ahead and select the **Tool Palette** option. Once selected, the **Subassembly Tool Palette** area will appear on the screen, as shown in *Figure 10.6*.

This tool palette essentially calls up all the potential subassemblies, or building blocks, that we can use to add to our overall assembly, which will eventually be used to model our roadway corridor later in this chapter:

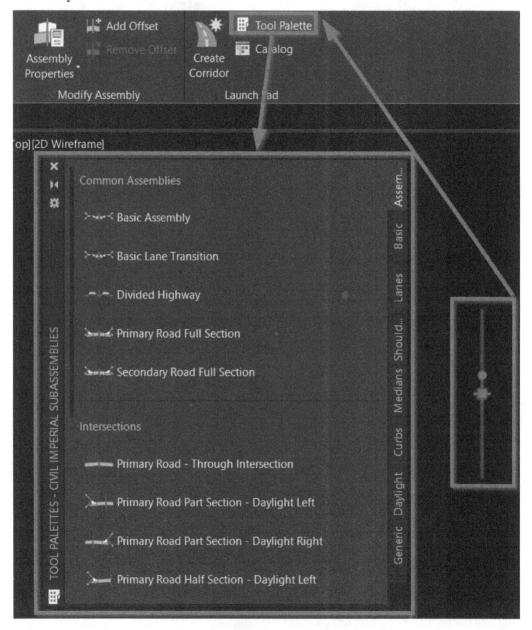

Figure 10.6 – Subassembly Tool Palette

Right off the bat, we will see that all subassemblies are displayed directly in the center section of the **Subassembly Tool Palette** area. Along the right-hand side of the **Subassembly Tool Palette** area, we will see a series of tabs that represent various groupings or categories of our subassemblies being displayed (refer to *Figure 10.7*).

Reading from top to bottom along the right-hand side, we have the following categories available to us:

- **Assemblies – Imperial**: Displays Civil 3D's stock assembly templates that have been pre-developed to include multiple subassembly objects

- **Basic**: Displays basic individual subassembly objects that we can use to custom-build our assemblies

- **Lanes**: Displays individual lane-focused subassembly objects that we can use to custom-build our assemblies

- **Shoulders**: Displays individual shoulder-focused subassembly objects that we can use to custom-build our assemblies

- **Medians**: Displays individual median-focused subassembly objects that we can use to custom-build our assemblies

- **Curbs**: Displays individual curb-focused subassembly objects that we can use to custom-build our assemblies

- **Daylight**: Displays individual daylight-focused subassembly objects that we can use to custom-build our assemblies

- **Generic**: Displays individual generic subassembly objects that we can use to custom-build our assemblies

- **Conditional**: Displays individual conditional subassembly objects that will allow us to automate decision-making as we further custom-build our assemblies:

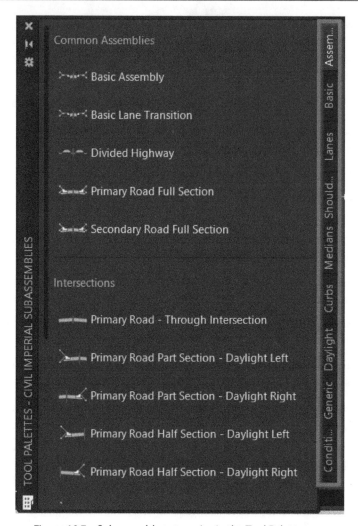

Figure 10.7 – Subassembly categories in the Tool Palette area

Furthermore, if we were to left-click in the bottom section of the tabbed area, we would be presented with a few more additional subassembly options that we can choose to be displayed in our **Subassembly Tool Palette** area (refer to *Figure 10.8*).

These additional subassembly tabs include the following:

- **Trench Pipes**: Displays Civil 3D's stock Trench Subassembly groupings, which have been pre-developed to include multiple subassembly objects

- **Retaining Walls**: Displays Civil 3D's stock Retaining Wall Subassembly groupings, which have been pre-developed to include multiple subassembly objects

- **Rehab**: Displays Civil 3D's stock Rehab Subassembly groupings, which have been pre-developed to include multiple subassembly objects

- **New Rehab**: Displays Civil 3D's stock New Rehab Subassembly groupings, which have been pre-developed to include multiple subassembly objects

- **Bridge**: Displays Civil 3D's stock Bridge Subassembly groupings, which have been pre-developed to include multiple subassembly objects.

- **Rail**: Displays Civil 3D's stock Rail Subassembly groupings, which have been pre-developed to include multiple subassembly objects:

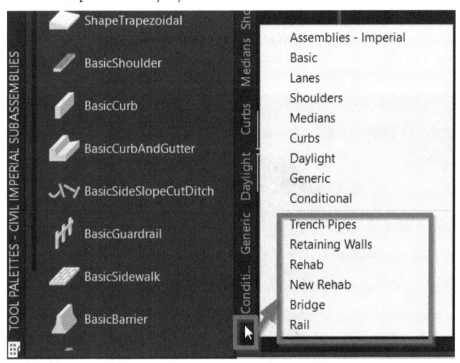

Figure 10.8 – Subassembly categories in the Tool Palette area (expanded)

To design our first Residential Subdivision, we're going to keep things fairly simple at this stage and leverage most of the basic and generic subassemblies available to us.

That said, let's select our **Basic** tab to display all Basic Subassembly objects available to us. Running through the following steps, we can begin creating our assembly by adding a **BasicLane** subassembly to our overall assembly (refer to *Figure 10.9* for visual steps):

1. Select the **BasicLane** subassembly.
2. Go to the **Properties** dialog box.

3. Fill in the **Parameters** values we intend to apply to the **BasicLane** subassembly. In this instance, we'll accept the default values, which are listed as follows:

    - **Version**: R2019

    - **Side**: Right

    - **Width**: 12.00'

    - **Depth**: 0.67'

    - **Slope**: -2.00%

4. We'll see a note on the command line stating: Select marker point within assembly or [Insert Replace Detached]:. At this point, we can simply select our assembly baseline in our model space since we are ultimately looking to build our assembly at this point in our design:

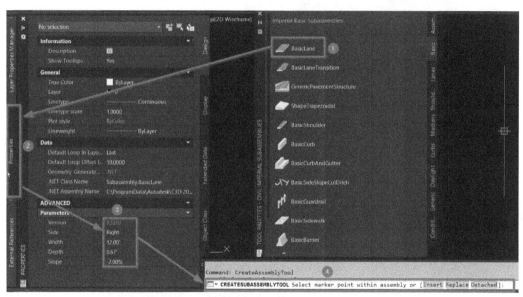

Figure 10.9 – Steps for adding a BasicLane subassembly

After selecting our assembly baseline, we will see our **BasicLane** subassembly attached to the centerline of our assembly. Before canceling out of the command, we can place our **BasicLane** subassembly as needed and change the values that are applied in the **Parameters** section of the **Properties** dialog as needed.

In our case, we will want to change the **Side** parameter from **Right** to **Left** and then select our assembly baseline again to place the left-hand side of our **BasicLane** subassembly, thereby completing a full road section. The new assembly should look similar to that shown in *Figure 10.10*:

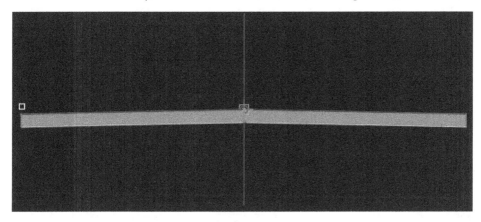

Figure 10.10 – BasicLane subassembly result

Next, we'll want to add the **BasicCurbAndGutter** subassembly to the ends of our **BasicLane** subassembly. Running through the following steps, we can continue building our assembly by adding a **BasicCurbAndGutter** subassembly to our overall assembly (refer to *Figure 10.11* for visual steps):

1. Select the **BasicCurbAndGutter** subassembly.

2. Go to the **Properties** dialog box.

3. Fill in the parameters we intend to apply to our **BasicCurbAndGutter** subassembly. In this instance, we'll accept the default values, which are listed as follows:

    -   **Version**: R2019

    -   **Side**: Left

    -   **Insertion Point**: Gutter Edge

    -   **Gutter Width**: 1.50

    -   **Gutter Slope**: -6.00%

    -   **Curb Height**: 0.75'

    -   **Gutter Height**: 0.50'

4. We'll see a note at the command line stating CREATESUBASSEMBLYTOOL: Select marker point within assembly or [Insert Replace Detached] :. At this point, we can simply select our top-left insertion node at the end of our **BasicLane** subassembly in our model space to attach the **BasicCurbAndGutter** subassembly at the edge:

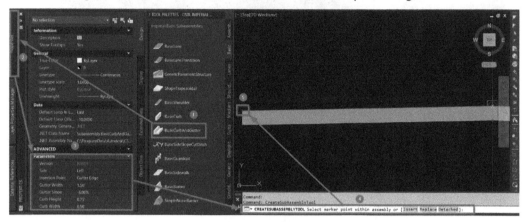

Figure 10.11 – Steps for adding a BasicCurbAndGutter subassembly

Now that we've placed the **BasicCurbAndGutter** subassembly on the left-hand side, let's cancel the command this time by hitting the *Esc* key on the keyboard. Although we still want to add the **BasicCurbAndGutter** subassembly to the right-hand side of our assembly, we'll take a slightly different approach and become more familiar with the **Subassembly Contextual** ribbon.

Follow these steps to add the **BasicCurbAndGutter** subassembly to the right-hand side of our assembly:

1. Select our newly placed **BasicCurbAndGutter** subassembly on the left-hand side of our assembly.

2. Go to the **Subassembly Contextual** ribbon and select the **Mirror** tool in the **Modify Subassembly** panel.

3. Select the top-right insertion node at the end of our **BasicLane** subassembly in our model space to attach the **BasicCurbAndGutter** subassembly at the edge:

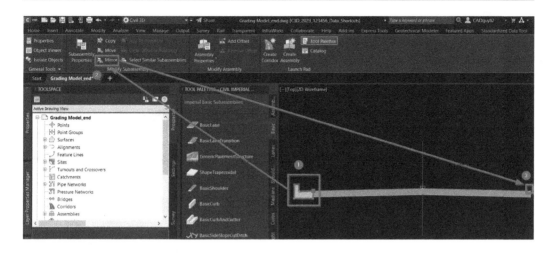

Figure 10.12 – Steps for mirroring a BasicCurbAndGutter subassembly

Although we can certainly continue building out our **ASM – Subdivision Main Road – Access** assembly by adding additional subassembly objects, we're going to temporarily put that on hold as we'll be looking at new ways to continue building our **ASM – Subdivision Main Road – Access** assembly later in this chapter.

Before jumping into the corridor creation process, we'll want to make one more assembly that will be used for our dead-end and cul-de-sac portions of the corridor. Since both instances will require parameters to be input that extend beyond our alignment geometry, we'll want to create an assembly that can run along a construction line representing our edge of pavement along our dead-end and cul-de-sac designs.

Instead of going back to the **Home** ribbon and running through the assembly creation process from scratch again, let's go ahead and start with the **ASM – Subdivision Main Road – Access** assembly we just created and make the necessary modifications for our dead-end and cul-de-sac designs.

That said, let's select the **ASM – Subdivision Main Road – Access** assembly baseline and use the AutoCAD COPY command (type COPY in the command line) and place the new assembly to the right, giving us two identical assemblies.

Let's focus our attention on the newly created assembly to the right and select all subassemblies and use the AutoCAD ERASE command (type ERASE in the command line) to remove all subassemblies attached to our new assembly.

Next, select our assembly baseline again, right-click with our mouse, and select the **Assembly Properties** option to pull up the **Assembly Properties** dialog box. Once the **Assembly Properties** dialog box has appeared on the screen, name it `ASM - Subdivision Main Road - Cul-De-Sac` and set **Description** to **Subdivision Cul-De-Sac Assembly**. Then, select the **OK** button toward the bottom of the dialog box (refer to *Figure 10.13* for the workflow):

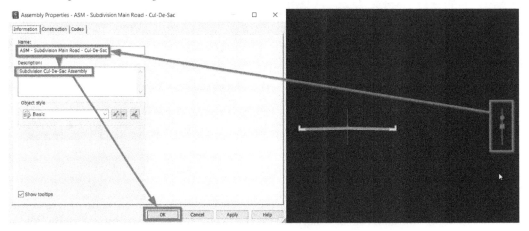

Figure 10.13 – Steps for modifying the Name and Description fields of our new assembly

To build our **ASM – Subdivision Main Road – Cul-De-Sac** assembly, we must add subassemblies to further build it out:

1. Select the **BasicCurbAndGutter** subassembly displayed on the right-hand side of our **ASM – Subdivision Main Road – Access** assembly.

2. Go to the **Subassembly Contextual** ribbon.

3. Navigate to the **Modify Subassembly** panel.

4. Select the **Copy** tool.

5. Select the baseline for our new **ASM – Subdivision Main Road – Cul-De-Sac** assembly.

6. Hit the *Esc* key twice on your keyboard to cancel the command and deselect all.

Next, we want to add a slightly different type of **Pavement** section that will allow us to target the centerline of our roadway alignments. The reason we are using this approach is to make sure that we can maintain the elevations we have already established in our Design Profiles.

If we were to apply the **BasicLane** subassembly, we would essentially end up with varying elevations that may be in conflict with what we are proposing. If any vertical adjustments are made later to our Design Profiles, we can be confident that the corridor models will automate dynamically with them as well.

Follow these steps to add a new subassembly that will target horizontal and vertical geometry to maintain the parameters set in our design (this workflow is shown in *Figure 10.14*):

1. In the **Tool Palette** area, select the **BasicLaneTransition** subassembly.

2. Activate the **Properties** dialog box.

3. Fill in the **Parameters** section as follows:

    *   **Version**: R2019

    *   **Side**: Left

    *   **Insertion Point**: Edge of Travel Way

    *   **Crown Point on Inside**: No

    *   **Default Width**: 12.00'

    *   **Depth**: 0.67'

    *   **Default Slope**: 2.00%

    *   **Transition**: Change offset and elevation

4. Select the top-left corner node of the **BasicCurbAndGutter** subassembly to insert and attach it to **BasicLaneTransition**:

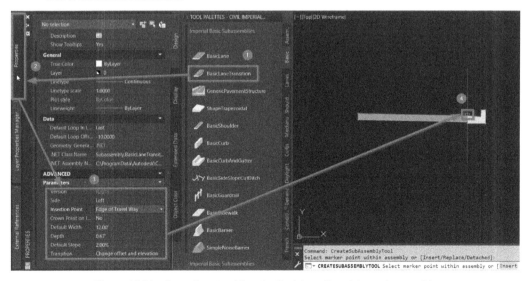

Figure 10.14 – Steps for attaching the BasicLaneTransition subassembly

Now that we have created our **ASM – Subdivision Main Road – Access** and **ASM – Subdivision Main Road – Cul-De-Sac** assemblies, let's learn how we can utilize them to create our corridor models.

Creating and modifying corridors

With our **ASM - Subdivision Main Road - Access** and **ASM - Subdivision Main Road - Cul-De-Sac** assemblies created, albeit not fully built out, let's learn how to create, manage, and manipulate our corridor models. So, let's create our first corridor model in our Residential Subdivision design.

Select the **ASM - Subdivision Main Road - Access** assembly baseline to activate the **Assembly Contextual** ribbon. Then, go to the **Launch Pad** panel and select the **Create Corridor** tool.

Once selected, we'll be presented with the **Create Corridor** dialog box, at which point we can fill in the fields as follows (also displayed in *Figure 10.15*), and then click on the **OK** button in the lower portion of the dialog box:

- **Name**: COR – Subdivision Main Road - Access
- **Description**: Subdivision access road from York Hwy
- **Corridor style**: Basic
- **Baseline type**: Alignment and profile
- **Alignment**: ALG – Subdivision Main Road – Access
- **Profile**: PRF – Subdivision Main Road – Access
- **Assembly**: ASM – Subdivision Main Road – Access
- **Target Surface**: <none>
- **Set baseline and region parameters**: *Uncheck the box*:

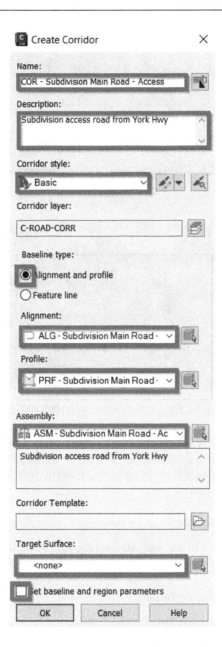

Figure 10.15 – The Create Corridor dialog box

After selecting the **OK** button in the **Create Corridor** dialog box, Civil 3D will run through the process of generating our new **COR - Subdivision Main Road - Access** corridor model along our **ALG – Subdivision Main Road – Access** alignment, with our Residential Subdivision design looking similar to that shown in *Figure 10.16*:

Figure 10.16 – Residential Subdivision for the first corridor model

Next, we'll want to run through the same steps to call up the **Create Corridor** dialog box again, but this time, we'll want to create our corridor along the **ALG – Subdivision Side Road – Cul-De-Sac** alignment. Once we've done that, we'll want to fill out the fields in the **Create Corridor** dialog box as follows (also displayed in *Figure 10.17*):

- **Name: COR – Subdivision Side Road – Cul-De-Sac**
- **Description: Subdivision Cul-De-Sac intersecting with Main Road**
- **Corridor style: Basic**
- **Baseline type: Alignment and profile**
- **Alignment: ALG – Subdivision Side Road – Cul-De-Sac**
- **Profile: PRF – Subdivision Side Road – Cul-De-Sac**
- **Assembly: ASM – Subdivision Main Road – Access**
- **Target Surface: <none>**
- **Set baseline and region parameters**: *Uncheck the box*:

Figure 10.17 – The Create Corridor dialog box

After selecting the **OK** button in the **Create Corridor** dialog box, Civil 3D will run through the process of generating the **COR – Subdivision Side Road – Cul-De-Sac** corridor model, with the updated Residential Subdivision design looking similar to that shown in *Figure 10.18*:

Figure 10.18 – Residential Subdivision for the second corridor model

As shown in our updated Residential Subdivision design, we have a couple of locations in one of the corridor models that will require some manipulation and updating. The following locations must be updated (also numbered and circled in *Figure 10.19*):

1. At the entrance of our Residential Subdivision design where our **ALG – Subdivision Main Road – Access** alignment intersects with our **ALG – Existing York Hwy – FromSurveyPoints** alignment, we'll want to move the starting point of our **COR – Subdivision Main Road – Access** corridor model to make room for our intersection design.

2. At the end of our **ALG – Subdivision Main Road – Access** alignment, we'll want to move the end point of our **COR – Subdivision Main Road – Access** corridor model to make room for our cul-de-sac design.

3. At the end of our **ALG – Subdivision Side Road – Cul-De-Sac** alignment, we'll want to move the end point of our **COR – Subdivision Side Road – Cul-De-Sac** corridor model to make room for our cul-de-sac design.

4. Where our **ALG - Subdivision Main Road – Access** alignment intersects with our **ALG - Subdivision Side Road - Cul-De-Sac** alignment, we'll want to make two adjustments. The first one will be to split our **COR – Subdivision Main Road – Access** corridor model and remove the section that intersects with our **ALG – Subdivision Side Road – Cul-De-Sac** alignment. The second will be to move the starting point of our **COR – Subdivision Side Road – Cul-De-Sac** corridor model. Both of these edits will essentially make room for our intersection design:

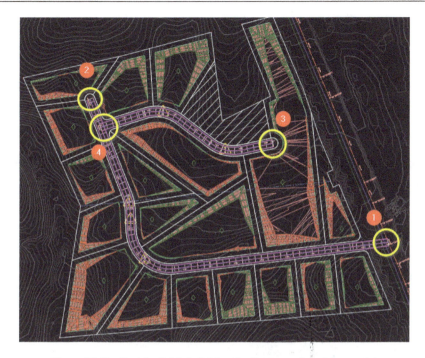

Figure 10.19 – Residential Subdivision for the second corridor model

To make edits *1-3* and part of *4*, we can simply select our corresponding corridor models, zoom into the beginning and end locations, select the appropriate grip with our mouse, and slide our corridor start and end points at approximately 50′, with the updated appearance looking similar to that shown in *Figure 10.20*:

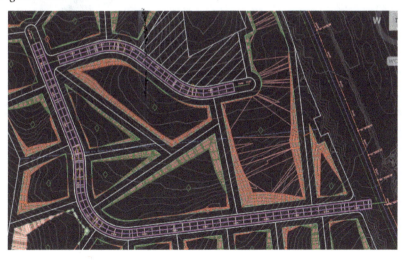

Figure 10.20 – Residential Subdivision after performing edits 1-3

To split our **COR – Subdivision Main Road – Access** corridor model and remove the section that intersects with our **ALG – Subdivision Side Road – Cul-De-Sac** alignment, we'll want to use the **Corridor Contextual** ribbon.

So, let's go ahead and select our **COR – Subdivision Main Road – Access** corridor model, then go to the **Modify Region** panel in the **Corridor Contextual** ribbon. We'll want to select the **Split Region** tool twice to create a section within our **COR – Subdivision Main Road – Access** corridor model that can be removed (refer to *Figure 10.21* for split locations):

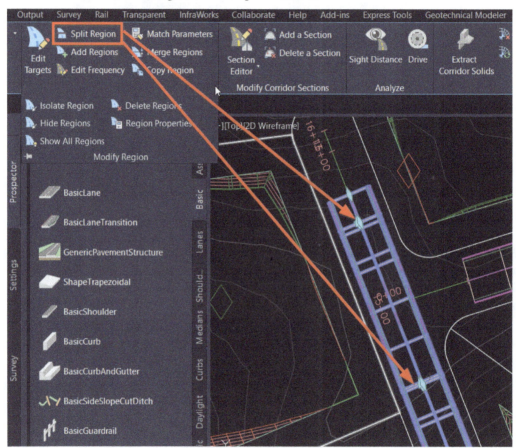

Figure 10.21 – Split Region for the corridor model

Next, we'll use the **Delete Regions** tool to remove the region that intersects with our **ALG – Subdivision Side Road – Cul-De-Sac** alignment (refer to *Figure 10.22*).

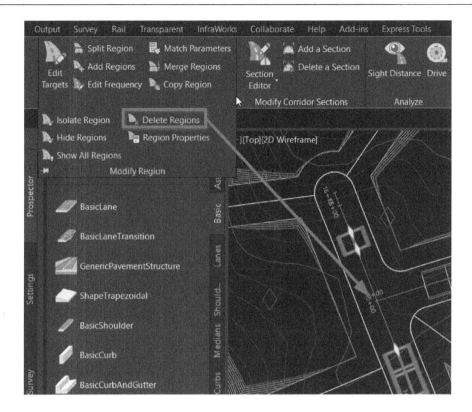

Figure 10.22 – Delete Regions for the corridor model

Now that we've made room for our intersection and cul-de-sac models along our alignments, let's go ahead and explore these tools a bit more to further our Roadway designs.

Creating and modifying intersections and cul-de-sacs

You may have noticed by now that our cul-de-sac situation needs to be updated a bit as it relates to our linework. As with most designs, there tends to be a bit of scope creep, and very rarely are we able to complete a design in our first pass.

In this instance, we had originally laid out our parcels with the assumption that the roadway following our **ALG – Subdivision Side Road – Cul-De-Sac** alignment would be a dead end. Our client has now asked that we make this an official cul-de-sac, which will require us to adjust our parcel layout slightly to accommodate the extended turnaround capacity.

That said, utilizing a combination of our AutoCAD tools and our Civil 3D tools, we'll need to make a few adjustments to our Residential Subdivision design. To update our Right-of-Way and parcels in the vicinity of our cul-de-sac, we'll want to start by opening up our `Site Plan Reference.dwg` file located within `Practical Autodesk Civil 3D 2024\Chapter 10\Reference.`

Once opened, let's zoom into the location of the cul-de-sac at the end of our **ALG – Subdivision Side Road – Cul-De-Sac** alignment, draw a circle with a 38' radius while the center of our circle is snapped to the end of our **ALG – Subdivision Side Road – Cul-De-Sac** alignment, then use the Trim command to trim out the circle portion that lies within the right-of-way, as shown in *Figure 10.23*:

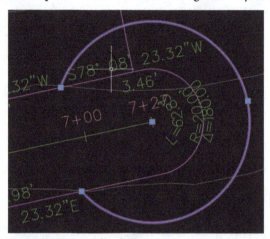

Figure 10.23 – Arc representing a new parcel boundary

Next, we'll go ahead and select our right-of-way parcel to activate our **Parcel Contextual** ribbon. Within the **Parcel Contextual** ribbon, we'll want to click on the **Edit Geometry** icon in the **Modify** panel, then select the **Edit Curve** tool within the **Edit Geometry** panel, as shown in *Figure 10.24*:

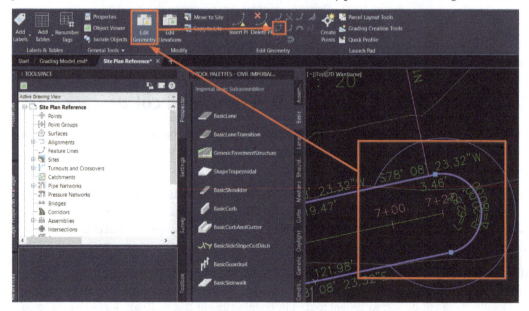

Figure 10.24 – Edit Curve parameters

Once the **Edit Curve** tool has been selected, at the command line, we'll be prompted to select the feature line curve to edit or delete, at which point we'll select the curve at the end of our right-of-way.

Once selected, the **Edit Feature Line Curve** dialog box will appear. Let's go ahead and change the **Radius** value to 38.00'. After changing the value and clicking the **Apply** button, we'll see the anticipated parcel layout automatically display (refer to *Figure 10.25*):

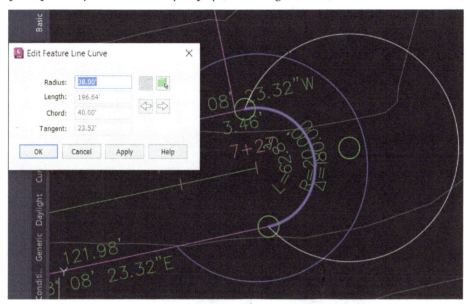

Figure 10.25 – Updated parcel curve layout

Then, click **OK** to accept and make a few more modifications. If we select the right-of-way parcel one more time, we'll want to move the grips to the end points of the temporary arc we created earlier (refer to *Figure 10.26* for the workflow):

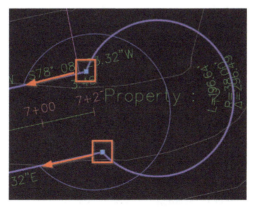

Figure 10.26 – Updated parcel layout

Once updated, go ahead and save your changes, and close out and reopen the `Grading Model.dwg` file located within `Practical Autodesk Civil 3D 2024\Chapter 10\Model`.

In the `Grading Model.dwg` file, do the following to create our edge-of-lane linework:

1. Zoom into the same location as before.

2. Draw a circle with a radius of **30 ft**.

3. Offset the **ALG – Subdivision Side Road – Cul-De-Sac** alignment by `12'` on either side (temporarily).

4. Use the offset lines to trim our newly created circle.

5. Fillet our offset lines and arc with a radius of `25'`.

6. Use the `Polyline Edit` or `Join` command to create one continuous polyline of all three arcs.

7. Erase the offset lines. We should be left with the edge-of-lane being displayed in our cul-de-sac, similar to that shown in *Figure 10.27*:

Figure 10.27 – Edge-of-lane arc in our cul-de-sac

The final steps we'll need to take before creating our cul-de-sac corridor model will be to convert our newly created arc into a feature line so that our cul-de-sac corridor model can use as a baseline while reading both horizontal and vertical geometry.

That said, follow these steps to accomplish this:

1. Select **Create Feature Line** from the **Object** option under **Home | Create Design | Feature Line**.

2. Select the arc depicting the edge-of-lane we just created and hit the *Enter* key on your keyboard.

3. When the **Create Feature Lines** dialog appears, we'll want to create a **New Site** in our first (top) field and name it **SIT – Proposed – Cul-De-Sac**. Then, set the **Description** property to **Cul-De-Sac Feature Lines** and select **OK**.

4. Change **Feature Line Style** to **Corridor Edge of Travel Way**.

5. Check the boxes for **Erase Existing Entities** and **Assign Elevations** and click the **OK** button.

6. In the **Assign Elevations** dialog box, we'll want to select the **From Surface** option to read our elevations from our **SRF – Existing Grade – FromSurveyPoints Surface** model and uncheck the box next to **Insert intermediate grade break points** so that we are left with one continuous slope through the entire length of our newly converted feature line.

> **Note**
>
> We will update elevation assignments later in this chapter.

With our feature line created, we can now use the same steps we went through in the previous section to create a corridor. When the `Create Corridor` command has been initiated, we'll want to fill in the corresponding fields in our **Create Corridor** dialog box:

- **Name**: COR – Subdivision Cul-De-Sac
- **Description**: Subdivision Cul-De-Sac
- **Corridor style**: Basic
- **Baseline type**: Feature Line
- **Site**: Cul-De-Sac
- **Feature Line**: FL – Subdivision Cul-De-Sac

We will likely need to use the **Select** option under **Drawing**. Once this icon is selected, we'll be prompted to physically select the feature line in our model space.

Once selected, we'll be prompted to input a name for that particular feature line.

Let's go ahead and name it **FL – Subdivision Cul-De-Sac** and click **OK**:

- **Assembly**: ASM – Subdivision Main Road – Cul-De-Sac
- **Target Surface**: <none>
- **Set Region and Baseline Parameters**: *Check the box*

Since we checked the box this time for **Set Region and Baseline Parameters**, we'll see the **Baseline and Region Parameters** dialog box appear. The main reason for pulling this dialog box up at this stage is to identify the target for the **BasicLaneTransition** subassembly so that it will target the horizontal and vertical geometry associated with our **ALG – Subdivision Side Road – Cul-De-Sac** alignment and our **PRF – Subdivision Side Road – Cul-De-Sac** profile, respectively.

With our **Baseline and Region Parameters** dialog box in view, let's go ahead and select the **Set all Targets** button (shown in *Figure 10.28*):

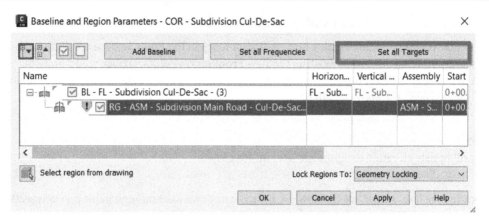

Figure 10.28 – The Baseline and Region Parameters dialog box

Then, select the **Transition** and **ALG – Subdivision Side Road – Cul-De-Sac** alignments (shown in *Figure 10.29*):

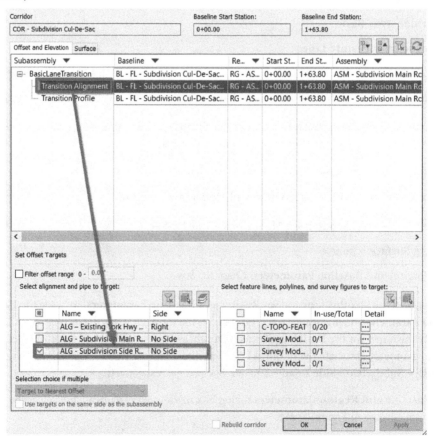

Figure 10.29 – Target Mapping | Transition Alignment

Then, select the **Transition** and **PRF – Subdivision Side Road – Cul-De-Sac** profiles (shown in *Figure 10.30*):

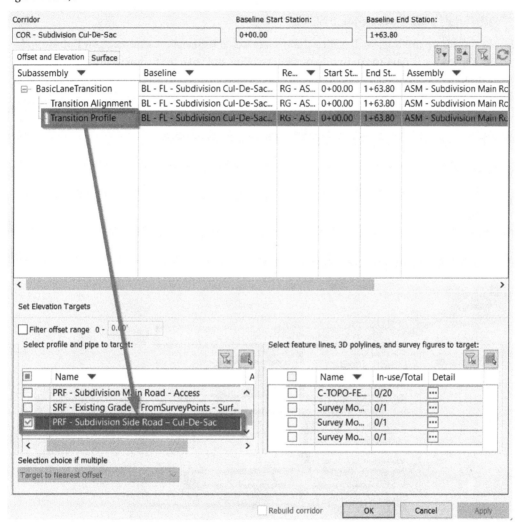

Figure 10.30 – Target Mapping | Transition Profile

Finally, select the **OK** button on both dialog boxes and let Civil 3D create the new cul-de-sac corridor model.

With our cul-de-sac created, let's move on to our intersection, where our **ALG – Subdivision Main Road – Access** and **ALG – Subdivision Side Road – Cul-De-Sac** alignments meet.

Let's follow these steps to familiarize ourselves with Civil 3D's Intersection Wizard:

1. Going back to **Home | Create Design**, we'll select the down arrow next to **Intersections** and select the **Create Intersection** tool.

2. In the command line, we'll notice that we are asked to select an **Intersection Point**. Once prompted, we'll want to zoom into our intersection and snap to the intersection point (using AutoCAD's OSNAPS) to identify the intersection point.

3. Civil 3D's **Create Intersection** dialog box will appear, starting with the **General** tab. We'll want to fill out the fields as follows in our **General** tab (also displayed in *Figure 10.31*) and select the **Next** button at the bottom of the **Create Intersection** dialog box:

   - **Intersection Name: INT – Subdivision Side Road – Cul-De-Sac**
   - **Description: Intersection of Subdivision Main Road and Side Road Alignments**
   - **Intersection marker style: Intersection Marker**
   - **Intersection label style: Basic**
   - **Intersection corridor type: All Crowns Maintained:**

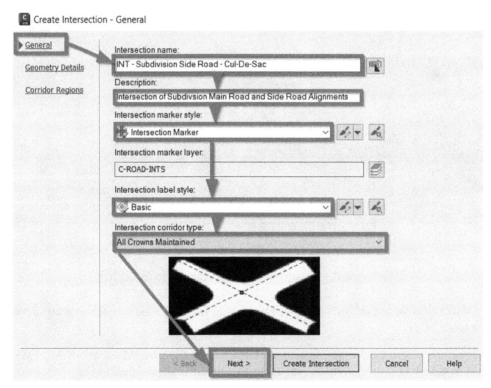

Figure 10.31 – Create Intersection | General

4. Next, we must fill in the fields in the **Geometry Details** tab as follows (also displayed in *Figure 10.32*) and select the **Next** button at the bottom of the **Create Intersection** dialog box:

    - **Create or specify offset alignments**: *Check the box*

    - **Create curb return alignments**: *Check the box*

    - **Create offset and curb return profiles**: *Check the box*:

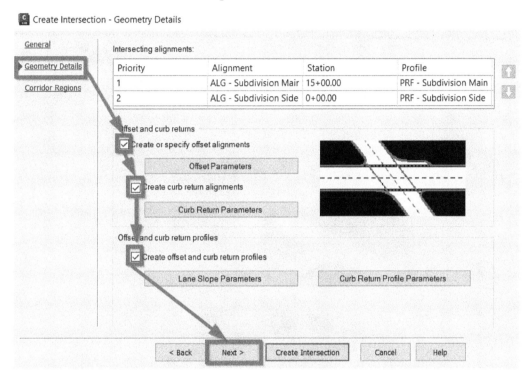

Figure 10.32 – Create Intersection | Geometry Details

5. Next, let's fill in the fields in the **Corridor Regions** tab as follows (also displayed in *Figure 10.33*) and select the **Create Intersection** button at the bottom of the **Create Intersection** dialog box:

    - **Create corridors in the intersection area**: *Check the box*

    - **Create a new corridor**: *Check the box*

- **Corridor Region Section Type: Accept all default Assemblies being applied:**

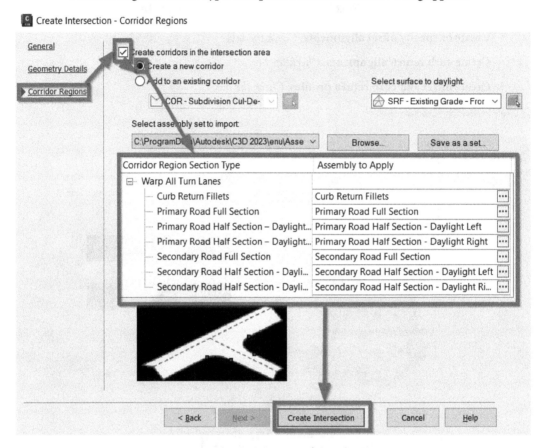

Figure 10.33 – Create Intersection | Corridor Regions

6. After selecting the **Create Intersection** button in the **Create Intersection** dialog box, our intersection corridor model will be created, similar to that shown in *Figure 10.34*:

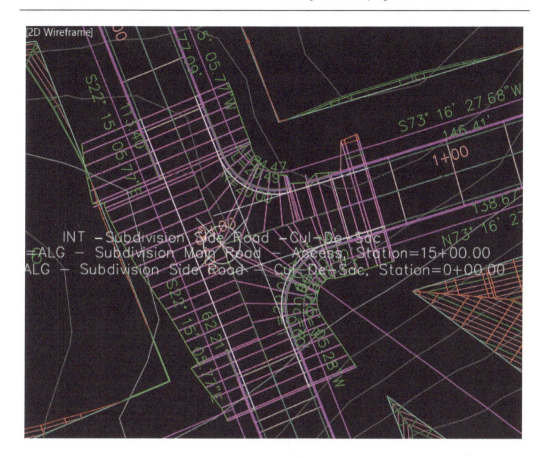

Figure 10.34 – Create Intersection | Corridor Regions

We'll notice a few things going on in our current model:

- The first, most obvious, one is that in this view alone (refer to *Figure 10.34*), a few additional subassemblies have been applied to our intersection corridor model.

- Second, if we navigate to the **Toolspace | Prospector** tab and expand our corridors and intersections, we'll notice that we now have a **Corridor - (1)** model. Although not very intuitive, this is associated with our newly created intersection.

  Let's go ahead and give this a name and description that is more closely aligned with its purpose in our Residential Subdivision design. If we right-click on the **Corridor – (1)** object and select the **Properties** option, we'll be presented with the **Corridor Properties** dialog box.

  If we activate the **Information** tab in the **Corridor Properties** dialog box, we'll want to set **Name** to COR - Intersection of Subdivision Main Road and Cul-De-Sac and **Description** to **Intersection of Subdivision Main Road and Side Road Alignments**, and then select the **OK** button in the lower portion of the **Corridor Properties** dialog box.

- The third item that has changed in our `Grading Model.dwg` file is that we now have a series of additional subassemblies that have been created and are available to us. If you recall back to the **Create Intersection** process, in the **Corridor Regions** tab, we accepted the defaults while listing out multiple assemblies to apply, at which point Civil 3D automatically generated and placed these assemblies into our model space (refer to *Figure 10.35*):

Figure 10.35 – Assemblies created during the intersection design process

As you can see with the default assemblies, there are quite a few differences between the subassemblies used to generate our **COR – Intersection of Subdivision Main Road and Cul-De-Sac** corridor model with those applied to all other corridor models in our file.

Using similar tactics to modify our assemblies in the previous section, we can use the **Assembly Contextual** ribbon, navigate to the **Modify Subassembly** panel, and use a combination of the **Copy**, **Move**, and **Mirror** tools available alongside the AutoCAD `Erase` command to rebuild our subassemblies. Ultimately, we'll want to replace the lanes applied in the default **Intersection Assemblies** area with a **BasicLane**, and then add our **Daylight Subassemblies** to our **Main Road Assemblies**.

It's also recommended to move the assemblies that were used to create our **COR – Intersection of Subdivision Main Road and Cul-De-Sac** corridor model so that they're closer to our other assemblies to ensure they are all grouped. After moving and applying all edits, our grouped assemblies should look similar to that shown in *Figure 10.36*:

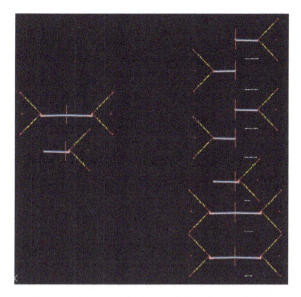

Figure 10.36 – Grouped assemblies after updating

Now that we've updated all our assemblies, we'll want to go back to the **Toolspace | Prospector** tab, scroll down to the corridor objects listed, right-click on **Corridors**, and select the **Rebuild All** option, as shown in *Figure 10.37*, allowing us to update all the models in our file in one clean swoop:

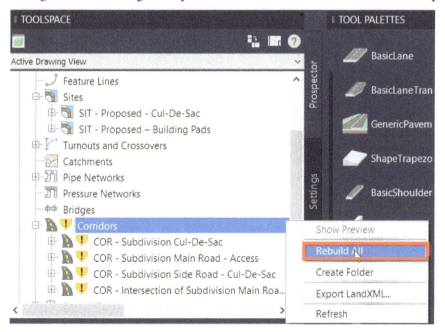

Figure 10.37 – Rebuild All

The final pieces to the corridor puzzle we'll want to perform are related to the targeting and frequency being applied to all of our corridors. At this point, our corridor models are essentially floating in the model space without any real tieback to our existing surface.

Now that we've added daylight subassemblies to all of our assemblies, we'll want to make sure we specify that we want to target our daylight lines to our **SRF – Existing Grade – FromSurveyPoints** surface model. Concerning the **Frequency** settings, we'll want to apply a tighter frequency to all of the corridor models in our file.

By applying a tighter frequency interval, we are ultimately making our roadway designs much more accurate and smoother, thus providing us with a better product in the end.

That said, let's go to the **Modify** ribbon toward the top of our Civil 3D session and select the **Corridor** tool within the **Design** panel, as shown in *Figure 10.38*:

Figure 10.38 – Modify | Design

Once the **Corridor** tool has been selected, our ribbon and panels will switch to the **Corridor Contextual** ribbon. If we shift our focus to the **Modify Region** panel, we'll be able to activate our **Edit Targets** tool, as well as our **Edit Frequency** tool, as shown in *Figure 10.39*:

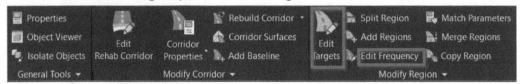

Figure 10.39 – Corridor Contextual | Modify Region

If we go ahead and select the **Edit Targets** tool, we'll be prompted to select the corridor that we'd like to update in the command line. After selecting one of our corridor models, we'll then be asked to specify the region within the selected corridor that we'd like to update, at which point we'll type (or select) **All**, bringing up the **Target Mapping** dialog box.

Now, let's use the **Select a surface for all surface targets** option and select our **SRF – Existing Grade – FromSurveyPoints** surface model, and then click the **OK** button at the bottom of our **Target Mapping** dialog box (refer to *Figure 10.40*).

Perform these steps to apply surface targeting to all additional corridor models in our current file:

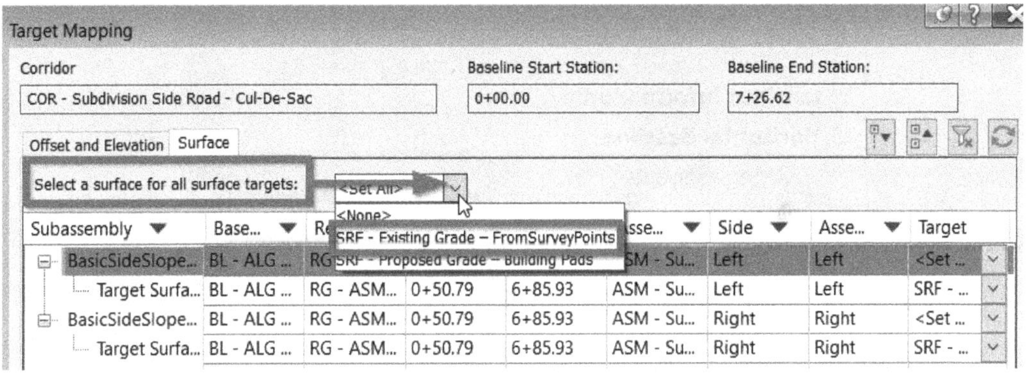

Figure 10.40 – The Target Mapping dialog box

Next, we'll want to update the frequency settings for all of our corridor models. Once the **Edit Frequency** tool has been selected, we'll be prompted to select the corridor that we'd like to update. After selecting any of our corridor models, we'll be prompted to select the region that we'd like to update.

Unlike the **Targeting** option, we will need to place our mouse within the corridor extents (corridor extents will be highlighted) and left-click to accept the identified section.

Once the section has been accepted, the **Frequency to Apply Assemblies** dialog box will appear, at which point we'll want to update the values to **5'** for now to create a much smoother and more accurate corridor model (refer to *Figure 10.41*).

> **Note**
>
> Changing our **Frequency** values to smaller intervals, although providing a much smoother and more accurate Corridor model, will also result in increased file size, longer Corridor rebuilding times, and longer opening and saving times.
>
> That said, at earlier stages of design, it is recommended that we use the default **Frequency** values. As we approach our final design or Issued for Construction phase, we'll want to apply smaller intervals to ensure that our design models are as accurate as we can get.

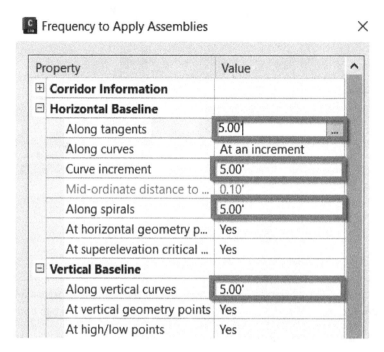

Figure 10.41 – The Frequency to Apply Assemblies dialog box

Now that we have created a fairly tight design for our corridor models, let's learn how to generate surfaces from our corridor models and integrate them into our overall Residential Subdivision design.

Creating a surface from corridors

With our corridor models created, we can now learn how to properly apply horizontal and vertical values associated with them to generate a dynamically linked surface model. As with all other modeled objects we've created, it's important to keep these links intact to minimize rework later on down the road and avoid the major risk of missing or overlooking a required update along the way.

If we go back to **Modify | Design** and select the **Corridor** tool again, we'll bring up the **Corridor Contextual** ribbon once more. There, we'll want to go to the **Modify Corridor** panel and select the **Corridor Surfaces** tool (refer to *Figure 10.42*):

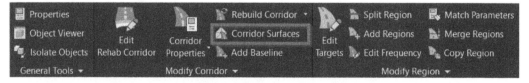

Figure 10.42 – Corridor Contextual | Modify Corridor | Corridor Surfaces

Once selected, we'll be prompted to select the corridor that we'd like to update. After selecting any of our corridor models, the **Corridor Surfaces** dialog box will appear. In the **Surfaces** tab, do the following to generate the surface model that will be dynamically linked to our corridor model (also shown in *Figure 10.43*):

1. Select the **Surfaces** tab.

2. Select the **Create a Corridor Surface** icon.

3. Set **Data type** input to **Links**.

4. Set **Specify code** to **Top**.

5. Select the **Add Surface Item** icon.

6. Set **Surface Style** to **Contours 1' and 5' (Design)**.

7. Check the **Add as Breakline** box (this setting will provide additional definition and accuracy for our surface model):

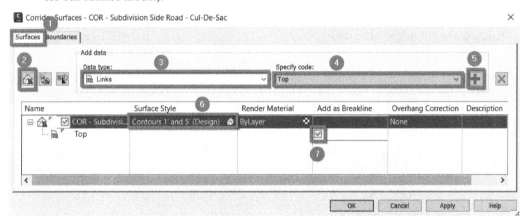

Figure 10.43 – Steps for creating a corridor surface

Next, we'll switch over to the **Boundaries** tab within the **Corridor Surfaces** dialog box, right-click on our newly created corridor surface model and select the **Corridor extents as outer boundary** option (refer to *Figure 10.44*), and then select the **OK** button at the bottom of the **Corridor Surfaces** dialog box.

By applying **Corridor extents as outer boundary** to our corridor surface model, we are ensuring that Civil 3D will not attempt to interpolate outside of the corridor limits and that our corridor surface boundary is terminating where it is meeting our **SRF – Existing Grade – FromSurveyPoints** surface model:

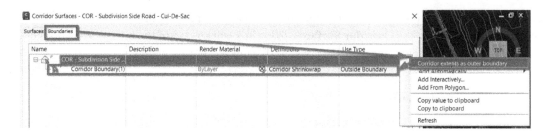

Figure 10.44 – Corridor Surfaces | Boundaries

After closing out of the **Corridor Surfaces** dialog box, a surface model will be displayed in our model space.

Using these same steps, let's go ahead and generate corridor surfaces for all four corridors. Once all four corridor surfaces have been created, we'll want to combine them into one comprehensive surface model:

1. Going back to the **Toolspace | Prospector** tab, we'll want to expand the surfaces to view all surface models in our current file.

2. After running through the process of creating a surface, as we learned in *Chapter 6, Surfaces - The First Foundational Component to Designs within Civil 3D*, name this comprehensive surface model SRF - Proposed Grade - Corridors with **Description** as **Combined Corridor Surface Models**.

3. Next, we'll want to expand **SRF – Proposed Grade – Corridors Surface**, then expand **Definitions**. If we right-click on **Edits** and select the **Paste Surface** option, we'll be presented with the **Select Surface to Paste** dialog box.

4. Here, we'll want to double-click on our surfaces in the following order (also shown in *Figure 10.45*):

   - **COR – Subdivision Main Road – Access Surface**

   - **COR – Subdivision Side Road – Cul-De-Sac Surface**

   - **COR – Intersection of Subdivision Main Road and Cul-De-Sac Surface**

   - **COR – Subdivision Cul-De-Sac Surface:**

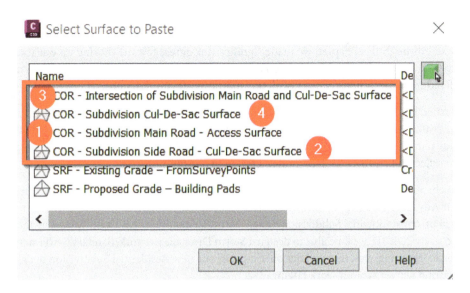

Figure 10.45 – The Select Surface to Paste dialog box

Once all the corridor surfaces have been pasted, resulting in one comprehensive **SRF – Proposed Grade – Corridors** surface model, we'll want to add to our **Data Shortcuts** project by performing the steps outlined in previous chapters using the **Create Data Shortcuts** option. The final output should look similar to that shown in *Figure 10.46*:

Figure 10.46 – Final Residential Subdivision appearance

Summary

As we worked through this chapter, we made significant progress toward shoring up our roadway design throughout our Residential Subdivision design. We learned why assemblies and subassemblies are considered to be the building blocks of all transportation-focused designs.

We also learned how to create corridor models, intersections, and cul-de-sacs using our alignments, profiles, assemblies, and subassemblies. We even sprinkled in some adjustments that needed to be made to our parcel layout due to design change requests, and how to create one combined surface model that reflects all corridor surface models created in our file.

In the next chapter, we'll take a few steps further beyond our typical Roadway Modeling tool belt and pull out our Advanced Roadway Modeling tool belt before progressing with our Utility modeling tool belt to finish up our Residential Subdivision design. By understanding what Utility Modeling tools are available to us in Civil 3D, we'll be able to design a Storm Drainage network, Sanitary Sewer network, and Domestic Water network that will not only support our new Residential Subdivision but also tie into the existing networks along York Highway.

Advanced Roadway Modeling Tool Belt for Everyday Use

In the previous two chapters, we had an opportunity to unwrap all of the land development and roadway modeling-focused tools available to us within Civil 3D, while also learning some very practical applications of how we can utilize them to further progress our residential subdivision design. Not only did we learn how to use these particular tools to create design objects within Civil 3D, but we also began to understand how best to manage these objects and the significance of each one to the overall design.

In this chapter, we'll start off by refining our proposed surface models a little more through corridor manipulation efforts to ensure that our residential subdivision will drain during rain events. Refinement of our proposed surface models will be necessary before we consider putting on our *Utility Modeling Tool Belt for Everyday Use* in *Chapter 12* and jump into the world of utility infrastructure that ultimately services our communities and land.

That said, in this chapter, we'll be covering the following topics:

- Updating assemblies and designing driveways
- Designing a dead end
- Designing our residential subdivision main entrance

With that, let's go ahead and open up our `Grading Model.dwg` file located within the `Practical Autodesk Civil 3D 2023\Chapter 11\Model` directory. Once opened, you'll notice that we'll be starting this chapter pretty much where we left off in the previous chapter, with the display of our model looking similar to that shown in *Figure 11.1*:

Figure 11.1 – Final appearance of our residential subdivision after Chapter 10

Technical requirements

The exercise files for this chapter are available at `https://packt.link/UoiPn`

Updating assemblies and designing driveways

With our `Grading Model.dwg` file open, you'll notice that there are a few items that need to be cleaned up and accounted for right off the bat:

- Side slopes in cut and fill locations have not been consistently applied throughout our site

- Driveways, or lot access, will need to be designed so that we can properly locate our drainage structure and driveway culvert locations throughout our site

- **SRF - Proposed Grade – Building Pads** and **SRF - Proposed Grade – Corridors** Surface models are overlapping, resulting in conflicting proposed elevations on our site

> **Note**
>
> This conflict has the potential to resolve itself after the first two updates are accounted for.

Tackling these updates in sequence, we'll start by zooming and panning over to all of our assemblies that were used to create all of our Corridor and Intersection models in *Chapter 10, Roadway Modeling Tool Belt for Everyday Use*. Once there, let's use the following steps to update the subassemblies used in our assemblies (also shown in *Figure 11.2*):

1. Go ahead and simply select all of the **BasicSideSlopeCutDitch** subassemblies on both the left and right sides of all of the assemblies.

2. Open up our **Properties** dialog box.

3. Update the slopes listed in our **Cut Slope**, **Fill Slope**, and **Foreslope Slope** fields to apply a 2:1 slope.

4. Update the slope listed in our **Backslope Slope** field to apply a 2:1 slope:

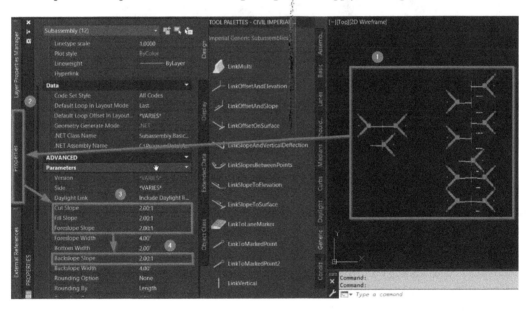

Figure 11.2 – Steps to update subassembly properties

5. From there, let's hop back over to our **Toolspace | Prospector** tab and scroll down to our **Corridors and Intersections** section. To update the corridors, we can simply right-click on the **Corridor** category and select the **Rebuild All** option. For our intersections, we'll need to expand the **Intersections** category, right-click on the **INT – Subdivision Side Road – Cul-De-Sac** intersection, and select the **Update Regions and Rebuild Corridor** option.

Lastly, we'll need to scroll up to and expand our **Surfaces** category, along with the **Proposed Conditions** subfolder, right-click on our **SRF - Proposed Grade – Corridors Surface** model, and select the **Rebuild** option. Once surfaces have been rebuilt, our roadway design should now be updated to accommodate the new slopes applied, with the next appearance of our site looking similar to that shown in *Figure 11.3*:

Figure 11.3 – Final appearance of our residential subdivision subassembly updates is applied

Although we still have a slight overlap of our proposed Surface models in the southeast corner of our site, we'll go ahead and add the driveway locations next and come back to this location when we perform a final cleanup of our Surface models.

With that, let's start adding our driveway locations to our residential subdivision layout. For this portion of edits, we'll start by adding our driveway locations into our `Grading Model.dwg` file, and then copy this geometry over to our `Site Plan Reference.dwg` file thereafter.

Note

We're starting in our `Grading Model.dwg` file to make sure that we have a good sense as to where we want to place our driveways based on the **SRF - Proposed Grade – Building Pads** Surface model. Our driveway linework will be essentially represented by polyline geometry, so we won't be able to data reference these objects from our `Grading Model.dwg` file, hence the reason for us needing to copy into our `Site Plan Reference.dwg` file so that this linework ends up on our final sheets.

If we were to measure the distance from our edge of pavement line, where our roadway meets the curb and gutter, all the way to our setback line, we'll measure out a distance of 28 ft. With our typical driveways being 12 ft wide, let's go ahead and create a polyline using basic AutoCAD workflows—that is, 12 ft wide by 28 ft long.

Once created, let's go ahead and place it on any one of our lots by using AutoCAD's move, copy, snap, and rotation commands to properly place it on all of our lots with a perpendicular alignment to our roadway geometry, with the next stage looking similar to that shown in *Figure 11.4*:

Figure 11.4 – Driveway locations added to the model

Once all driveway locations have been added, let's go ahead and select all polylines representing the outlines of the driveways and use AutoCAD commands to copy and paste over into our `Site Plan Reference.dwg` file located in the `Practical Autodesk Civil 3D 2023\Chapter 11\Reference` directory. After all polylines have been copied over, we'll go ahead and place them on the **C-ROAD-FEAT** layer, save, and close out of our `Site Plan Reference.dwg` file.

Now, back in our `Grading Model.dwg` file, we'll jump back over to our assemblies and make a couple more modifications that will be applied to all. Ultimately, all we are going to do at this stage is add some **ConditionalHorizontalTargeting** subassemblies in between our **BasicLane** and **BasicCurbAndGutter** subassemblies.

By adding the **ConditionalHorizontalTargeting** subassemblies, we are adding instances where our Civil 3D Corridor and Intersection models will have to make a decision. In the event that driveway polylines are being detected perpendicular to the baseline alignments, we want our corridor to model out a portion of the anticipated driveway into each parcel. In the event that there is no driveway polyline detected, our corridor will continue the design as is.

With that, let's activate our **Subassembly** Tool Palette again and use the following steps to update our assemblies:

1. Activate the **Conditional** tab in our **Subassembly** Tool Palette.

2. Select the **ConditionalHorizontalTarget** subassembly.

3. Fill out the parameters in the **Properties** dialog box, as follows:

    - **Side**: **Left**

    - **Layout Width**: 12.00'

    - **Layout Grade**: 1.00:1

    - **Type**: **Target Not Found**

    - **Maximum Distance**: 9999.00'

4. Select the left side of the **BasicLane** subassembly.

5. With our **ConditionalHorizontalTarget** subassembly still activated, go back to the **Parameters** section in the **Properties** dialog box.

6. Switch the **Type** value from **Target Not Found** to **Found**.

7. Select the left side of the **BasicLane** subassembly.

8. Cancel out of the insertion process by hitting the *Esc* key on your keyboard, with the resulting assembly appearing similar to that displayed in *Figure 11.5*:

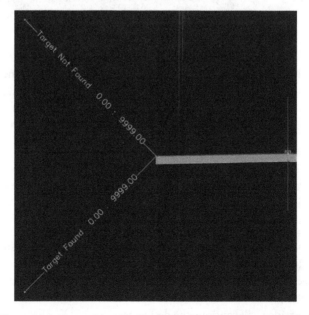

Figure 11.5 – Assembly with ConditionalHorizontalTargets added

9. Select the **BasicCurbAndGutter** and **BasicSideSlopeCutDitch** subassemblies on the left side of the assembly to activate the subassemblies' contextual ribbon.

10. Select the **Move** tool in the **Modify Subassembly** panel.

11. Move the **BasicCurbAndGutter** and **BasicSideSlopeCutDitch** subassemblies to the **ConditionalHorizontalTarget** subassembly with **Target Not Found** defined as the **Type** value.

With our **Target Not Found** type all configured, let's jump down to focus on our **Found** type to place our **Driveway** and **Daylight** subassemblies on our **SRF - Proposed Grade – Building Pads** Surface model. We'll use the following steps to accomplish this:

1. Select our **BasicLane** subassembly to activate our subassembly contextual ribbon.

2. Using the **Copy** command in the **Modify Subassembly** panel, we'll copy the **BasicLane** subassembly to the found **ConditionalHorizontalTarget** subassembly on the same side of our assembly.

3. Hit the *Esc* key to deselect all objects.

4. Select the newly copied **BasicLane** subassembly at the end of the found **ConditionalHorizontalTarget** subassembly and go over to our **Properties** dialog box.

5. Going down to the **Parameters** section, we'll change the following values:

    *   **Width**: 28.00'

    *   **Slope**: 0.05%

6. Hit the *Esc* key to deselect all objects.

7. Select the **BasicSideSlopeCutDitch** subassembly on the same side of the assembly that we've been working on located at the end of the not found **ConditionalHorizontalTarget** subassembly to activate the subassembly contextual ribbon again.

8. Using the **Copy** command in the **Modify Subassembly** panel, we'll copy the **BasicSideSlopeCutDitch** subassembly to the end of the newly copied **BasicLane** subassembly, with the final output looking similar to that shown in *Figure 11.6*:

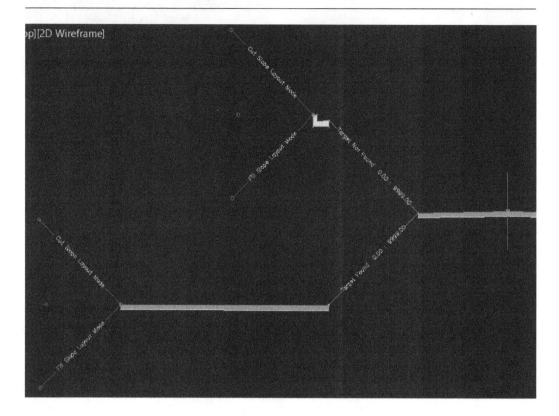

Figure 11.6 – Updated subassembly

Now that we've updated the left side of one of our assemblies, we'll need to apply this configuration to the remainder of our subassemblies. You can either run through the same steps as outlined earlier to modify the remaining assemblies, switching from the **Left** to the **Right** side in the **Parameters** section in the **Properties** dialog box as required.

Or, you can simply use the subassembly contextual ribbon to mirror, copy, and move this configuration to all remaining assemblies, similar to the steps outlined in *Chapter 10, Roadway Modeling Tool Belt for Everyday Use*. Either way, the final assembly configurations should look similar to that shown in *Figure 11.7*:

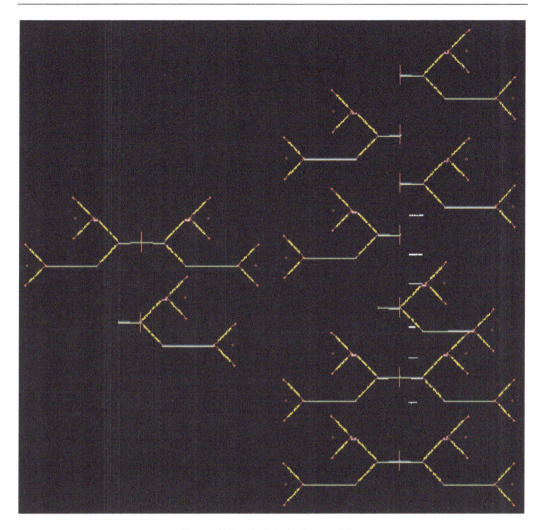

Figure 11.7 – Updated subassemblies

With all of our assemblies updated now, the next step we'll want to take is to update our corridors and intersections. If we go back to our **Toolspace | Prospector**, let's scroll down to our **Corridor** objects and expand the category. Starting with our **COR - Subdivision Main Road - Access Corridor** model, let's go ahead and right-click on it and select the **Properties** option. When our **Corridor Properties** dialog box appears, we'll want to switch over to the **Parameters** tab so that we can specify the new **Driveway** objects and our **SRF - Proposed Grade – Building Pads** Surface model to target all driveway locations.

In our **Parameters** tab, let's go ahead and select the ellipsis symbol at the top of the **Target** column so that we can apply similar targets across the entire length of our **COR - Subdivision Main Road - Access Corridor** model—refer to *Figure 11.8*:

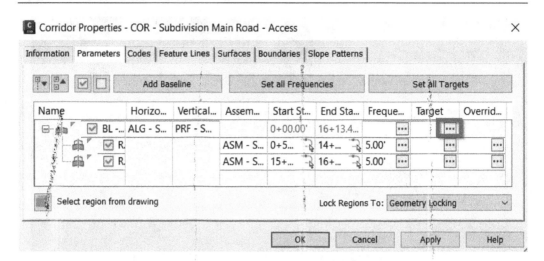

Figure 11.8 – Applying targets to all regions within a corridor

Once selected, a **Target Mapping** dialog box will then appear, allowing us to apply **Offset**, **Elevation**, and **Surface** targets to all subassemblies that have the option to target **Feature Lines**, **Polylines**, **Survey Figures**, **Alignments**, **Pipes**, **Profiles**, and **Surface** objects within our file(s).

In the **Offset and Elevation** tab, we'll want to use the following steps to apply horizontal targets so that our corridor models will be able to intelligently make a decision within our file on how we should be grading our site based on the parameters we are feeding into it (also shown in *Figure 11.9*):

1. Select the **Offset and Elevation** tab.

2. Select all target offsets listed out.

3. Click **Select** from the drawing option and select all polylines representing the driveways from the Site Plan Reference.dwg file along the corresponding Corridor model.

4. Check the **Use targets on the same side as the subassembly** box:

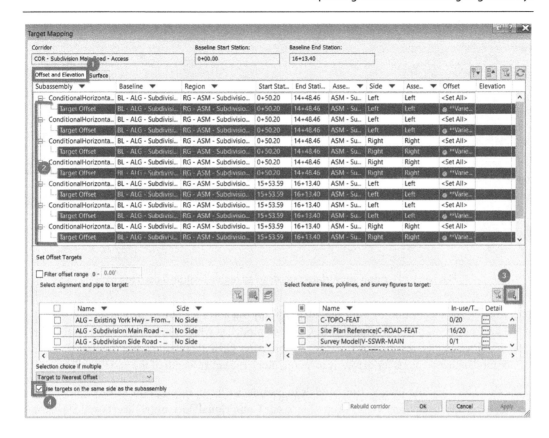

Figure 11.9 – Applying offset and elevation targets to all regions within a corridor

Continuing on, we'll use the following steps to finish up our targeting (also shown in *Figure 11.10*):

1. Select the **Surface** tab.

2. Select all target surfaces and select the corresponding surface to target inbound **Found** and **Not Found** situations.

3. Select the **OK** button:

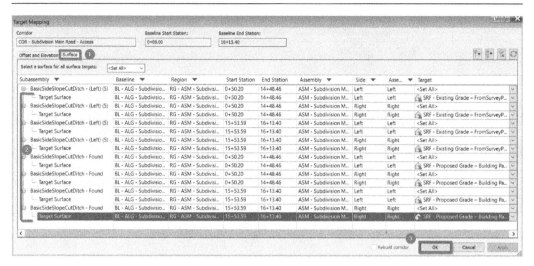

Figure 11.10 – Applying surface targets to all regions within a corridor

After selecting the **OK** button in both the **Target Mapping** and **Corridor Properties** dialog box, you will be asked if you'd like to rebuild your Corridor model, at which point we'll want to select the option that does, in fact, rebuild our Corridor model. After Civil 3D rebuilds, you'll notice that our Corridor model now accounts for driveway locations along the entire length of the baseline alignment.

For all other Corridor and Intersection models in our file, we'll run through the same steps we just outlined to apply horizontal and vertical targeting to our **ConditionalHorizontalTargets** and **Daylight** subassemblies used to rebuild all remaining Corridor models.

Corridors have incredible flexibility and intelligence that, when applied the proper way, can make the design seamless. In this section, we learned some vital terms for understanding how corridors work, and in the next section, we will look at another portion of corridors—designing common conditions seen in roadways such as a dead end.

Designing a dead end

For our dead-end design at the end of our **COR - Subdivision Main Road – Access** Corridor model, we need to take a slightly different approach. Since a dead-end design is essentially the end of the road, and since we have applied a curb and gutter throughout the entire residential subdivision design, we'll want to make a new assembly that includes just the curb and gutter along with the **ConditionalHorizontalTargeting** and **Daylight** options we included in the preceding section.

Before doing so, though, we'll need to create a new feature line that will contain both **Horizontal** and **Vertical** values for our new dead-end Corridor model to use as a baseline. With that, let's move to our **Feature Line** tools, which are located in the **Create Design** panel within the **Home** ribbon. If we select the down arrow next to **Feature Lines**, we'll then want to select the **Create Feature Line** tool (refer to *Figure 11.11*):

Figure 11.11 – Creating a feature line for a dead-end design

Once selected, we'll be prompted with a **Create Feature Lines** dialog box, at which point we'll want to create a new site, fill out the dialog box as follows (also shown in *Figure 11.12*), and then click on the **OK** button at the bottom of the dialog box:

- **Site**: **SIT – Proposed – Dead End**
- **Style**: **Corridor Edge of Paved Shoulder**

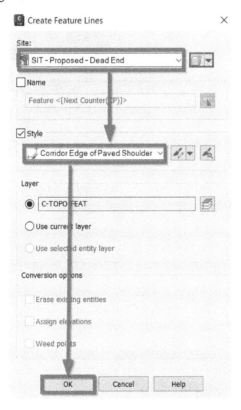

Figure 11.12 – Create Feature Lines dialog box

After selecting the **OK** button, we'll be prompted at the command line to specify a start point. Starting with the northeast portion of our dead end, we'll snap to the endpoint where our daylight line is.

Once clicked, we'll then be prompted to specify elevation or [**Surface**], at which point we'll type S for surface and select our **SRF – Proposed Grade – Corridors** Surface model, and click **OK.** This way, every point we click along the edge of our **COR - Subdivision Main Road – Access** Corridor model will be reflective of the elevations generated from our **SRF – Proposed Grade – Corridors** Surface model. We'll then select the points in sequential order, as shown in *Figure 11.13*:

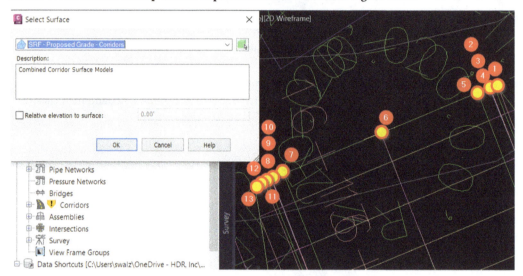

Figure 11.13 – Create Feature Line steps

With our feature line created and reading the elevations from our **COR - Subdivision Main Road – Access** Corridor model, let's jump over to our assemblies again.

Looking at all of our assemblies created thus far, we can quickly determine that there is one assembly already created that most closely resembles the dead-end assembly that we need to apply at this location in our design.

You guessed it! It's the **Curb Return Fillets** assembly that was automatically generated during our intersection design in *Chapter 10, Roadway Modeling Tool Belt for Everyday Use.*

Let's go ahead and select the **Curb Return Fillets Assembly** baseline and initiate the AutoCAD **Copy** command to create a duplicate version of this assembly and place it on the right side of all assemblies displayed.

Once copied, we'll want to make two modifications to this new assembly. The first modification will be to rename the assembly by selecting the baseline, pulling up our **Properties** dialog box, and changing the name from **Curb Return Fillets** to **Dead End**.

The second modification we'll want to make is to remove the **BasicLaneTransition** subassembly attached to the left side of the assembly, with the final version appearing similar to that shown in *Figure 11.14*:

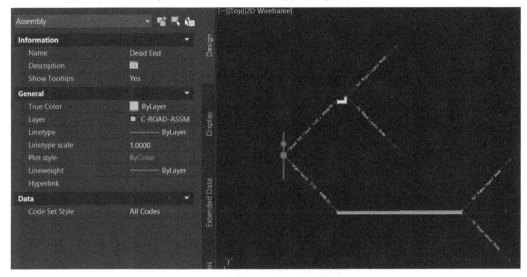

Figure 11.14 – Dead End subassembly

The final step we'll take here will be to create a new Corridor model specific to our dead-end design. After selecting our **Create Corridor** tool within the **Create Design** panel in our **Home** ribbon, we'll fill out the **Create Corridor** dialog box that appears as follows (also displayed in *Figure 11.15*) and click the **OK** button:

- **Name**: COR - Subdivision Main Road - Dead End

- **Description**: Dead End design at end of Main Road

- **Corridor style**: Basic

- **Baseline type**: Feature line

- **Site**: SIT – Proposed – Dead End

- **Feature line**: Dead End (we'll need to click the box icon to select the new feature line we just created, at which point we'll then be prompted to name said feature line, giving this Feature Line the name of **Dead End**)

- **Assembly**: Dead End

- **Target Surface**: <none>

- **Set baseline and region parameters**: *Check the box*

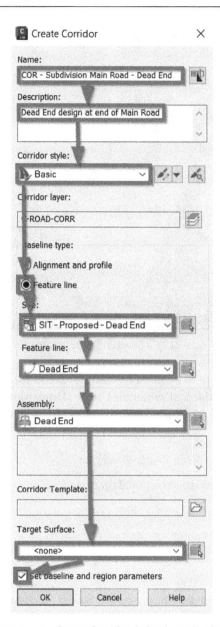

Figure 11.15 – Create Corridor dialog box (dead end)

Since we checked the box next to **Set baseline and region parameters** in our **Create Corridor** dialog box, a **Baseline and Regions parameters** dialog box will then appear. Using similar processes outlined in the previous section, we'll want to apply horizontal and surface targeting to our **COR - Subdivision Main Road - Dead End** Corridor model as we had to all other Corridor models in our Grading Model.dwg file, with the final output looking similar to that shown in *Figure 11.16*:

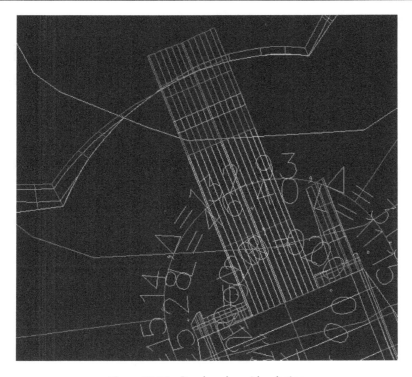

Figure 11.16 – Dead-end corridor design

With that, we will conclude our example of creating a dead-end corridor using the several different methods available to us in Civil 3D. Corridors come in many different forms, and the patterns and methods used to create one may not work for another.

Civil 3D allows for many different methods and intelligent tools to allow the creation of any corridor design needed. In the next section, we will dive into the design of an entrance.

Designing our residential subdivision main entrance

With the final piece of the corridor modeling to go, let's jump over to our main entrance where our residential subdivision design meets with York Highway. As outlined in *Chapter 10, Roadway Modeling Tool Belt for Everyday Use*, to model out the intersection of our **ALG - Subdivision Main Road – Access** and **ALG - Subdivision Side Road - Cul-De-Sac** alignments, we'll apply similar steps to design our intersection between our **ALG – Existing York Hwy – FromSurveyPoints** and **ALG - Subdivision Main Road – Access** alignments.

Going back up to our **Create Design** panel inside of the **Home** ribbon, we'll activate the **Create Intersection** tool. Once activated, we'll be asked at the command line to identify the intersection point where we'll want to snap to the intersection of our **ALG – Existing York Hwy – FromSurveyPoints** and **ALG - Subdivision Main Road – Access** alignments, as shown in *Figure 11.17*:

Figure 11.17 – Intersection design location

After clicking on the intersection point, Civil 3D's **Create Intersection** dialog box will appear, starting with the **General** tab. We'll want to fill out the fields as follows in our **General** tab (also displayed in *Figure 11.18*) and select the **Next** button at the bottom of the **Create Intersection** dialog box:

- **Intersection name: INT – Subdivision Main Road – Entrance**
- **Description: Intersection of Subdivision Main Road and York Highway Alignments**
- **Intersection marker style: Intersection Marker**
- **Intersection label style: Basic**
- **Intersection corridor type: Primary Road Crown Maintained**

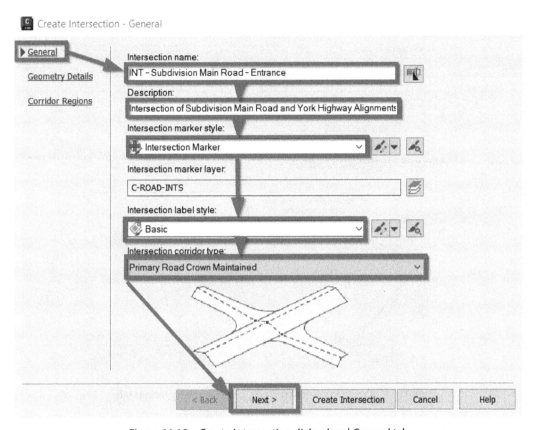

Figure 11.18 – Create Intersection dialog box | General tab

Next, we'll fill out the fields in the **Geometry Details** tab as follows (also displayed in *Figure 11.19*), but will also make some slight adjustments to our **Curb Return** parameters:

- **Create or specify offset alignments**: *Check the box*
- **Create curb return alignments**: *Check the box*
- **Create offset and curb return profiles**: *Check the box*

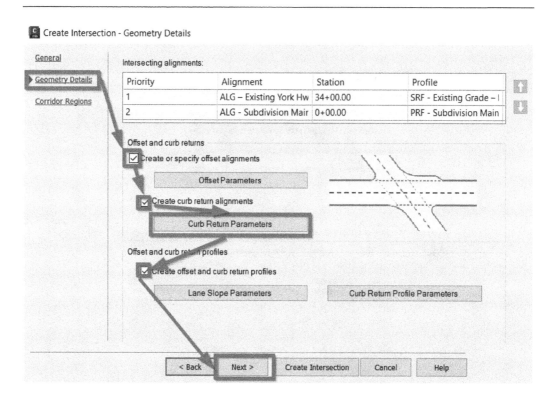

Figure 11.19 – Create Intersection dialog box | Geometry Details tab

Before selecting **Next**, we'll want to make sure we select the **Curb Return** parameters option so that we can make some adjustments to the radii being applied at the entrance of our residential subdivision.

Once selected, we'll see that the **Intersection Curb Return Parameters** dialog box appears, at which point we'll want to change our default radius value from 25.00' to 50.00' for both **SW - Quadrant** and **NW - Quadrant**.

You may also notice a preview of where the currently selected quadrant will be applied within our design by showing a preview in Model Space (refer to *Figure 11.20*):

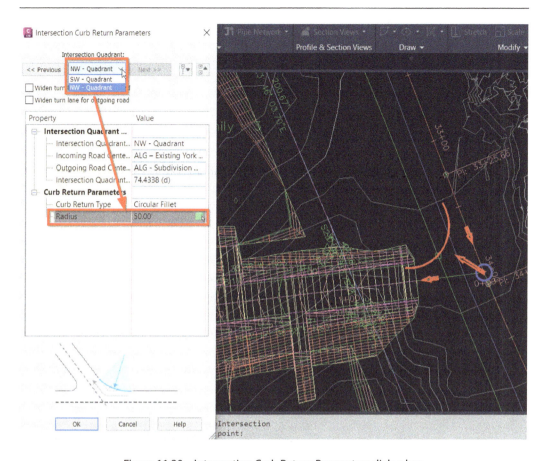

Figure 11.20 – Intersection Curb Return Parameters dialog box

After updating our radius values to both quadrants, we'll then select the **OK** button in our **Intersection Curb Return Parameters** dialog box and then click the **Next** button back in our **Create Intersection** dialog box within the **Geometry Details** tab, bringing us to our final tab of **Corridor Regions**.

Within the **Corridor Regions** tab, we'll fill out the values as follows (also displayed in *Figure 11.21*) and select the **Create Intersection** button at the bottom of the **Create Intersection** dialog box:

- **Create corridors in the intersection area**: *Check the box*
- **Create a new corridor**: *Check the box*
- **Corridor Region Section Type**: *Accept all default assemblies being applied*

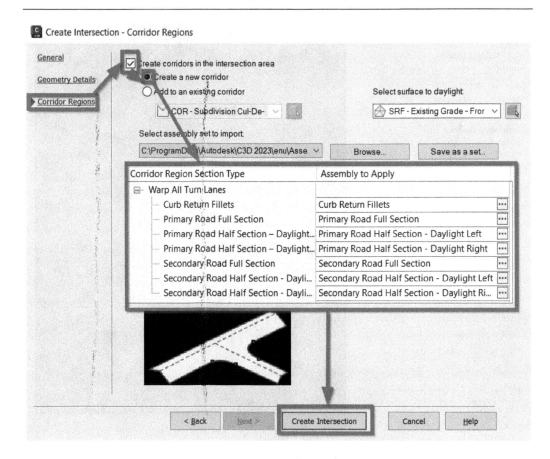

Figure 11.21 – Create Intersection dialog box | Corridor Regions tab

After selecting the **Create Intersection** button in the **Create Intersection** dialog box, our intersection Corridor model will be created similarly to that shown in *Figure 11.22*:

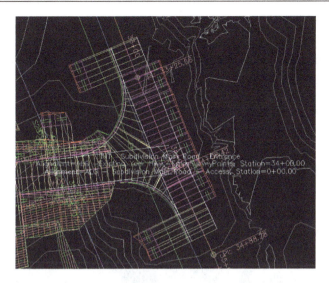

Figure 11.22 – Intersection design output at the main entrance

We'll notice that in this view alone (refer to *Figure 11.23*), we have a bit of overlap going on with our new **INT – Subdivision Main Road – Entrance** Intersection model and our **COR - Subdivision Main Road - Access** Corridor model.

Let's go ahead and select our **COR - Subdivision Main Road - Access** Corridor model and manually adjust the starting point by sliding our grip to the end of the **INT – Subdivision Main Road – Entrance** Intersection model, as shown in *Figure 11.23*:

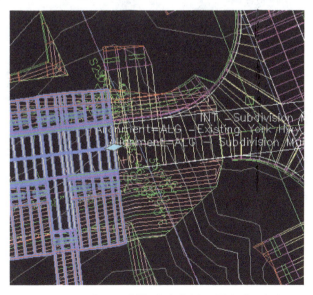

Figure 11.23 – New starting point for our COR - Subdivision Main Road - Access Corridor model

The second portion of the **INT – Subdivision Main Road – Entrance** Intersection model that will require a bit of cleanup involves the **ConditionalHorizontalTargets** subassembly. Now that we've essentially gone in and updated all of our assemblies to include the **ConditionalHorizontalTarget** subassembly, including those used to generate our Intersection models, we'll need to make sure that at least one target is defined in order for the daylighting to properly finish throughout the **INT – Subdivision Main Road – Entrance** Intersection model.

Since no actual driveways are placed within the extent of our **INT – Subdivision Main Road – Entrance** Intersection model, we'll want to make sure that we select only one driveway polyline that is not perpendicular to any of our anticipated projections.

That said, we may need to select a couple of different driveway polylines based on the anticipated horizontal projection to make sure that our subassemblies follow the **Not Found** application of our assemblies. Once all have been identified and we update our **INT – Subdivision Main Road – Entrance** Intersection model, we should have our design appear similar to that shown in *Figure 11.24*:

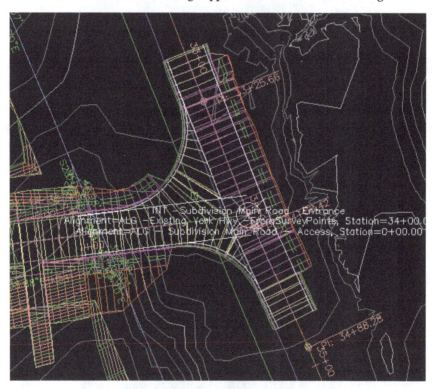

Figure 11.24 – Final intersection design output at the main entrance

With all of our Corridor and Intersection models updated and refined, we'll now want to create Corridor Surface models from our Corridor models by running through the steps outlined in *Chapter 10, Roadway Modeling Tool Belt for Everyday Use.*

Once all Corridor Surface models have been created, we'll then want to update our **SRF - Proposed Grade – Corridors** Surface model by pasting all remaining Corridor Surface models in and rebuilding our Surface model, with the final result looking similar to that shown in *Figure 11.25*:

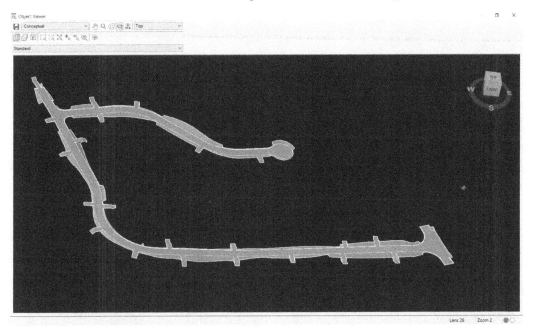

Figure 11.25 – Final SRF - Proposed Grade – Corridors Surface model

To complete our Surface model within our residential subdivision design, we'll now want to merge our **SRF - Proposed Grade - Corridors** and our **SRF - Proposed Grade – Building Pads** Surface models into one comprehensive Surface model so that we can accurately design our drainage utility network in *Chapter 12, Utility Modeling Tool Belt for Everyday Use.*

That said, let's go ahead and create a new surface using similar steps outlined in *Chapter 6, Chapter 6, Surfaces - The First Foundational Component to Designs within Civil 3D,* and *Chapter 9, Land Development Tool Belt for Everyday Use,* and place our new Surface model within our Proposed subfolder within the **Surfaces** category inside of our **Toolspace | Prospector** tab.

When creating our new surface, we'll give it a **Name** value of **SRF - Proposed Grade - Residential Subdivision** and a **Description** value of **Combined Corridor and Building Pads Surface Models**. After it's created, we'll paste in both our **SRF - Proposed Grade - Corridors** and our **SRF - Proposed Grade – Building Pads** Surface models to create one comprehensive Surface model, with the final output looking similar to that displayed in *Figure 11.26*. The order in which we paste each is important in this example. First, we will paste our building pad surface and then the corridors:

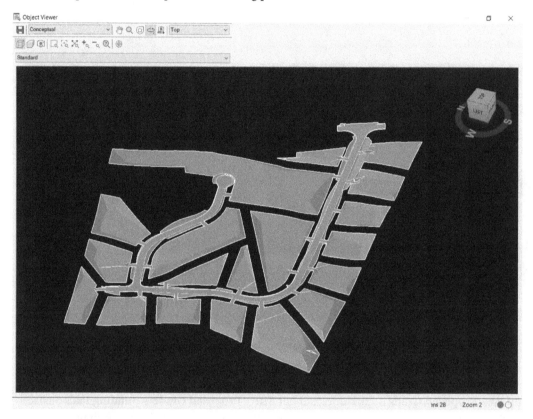

Figure 11.26 – Final SRF - Proposed Grade - Residential Subdivision Surface model

The last step we'll take is to create a data shortcut of our **SRF - Proposed Grade - Residential Subdivision** Surface model. We'll want to use this new Surface model to base our drainage utility design on in *Chapter 12, Utility Modeling Tool Belt for Everyday Use*, to ensure that we are properly draining our residential subdivision design during rain events.

Summary

As we worked through this chapter, we have been able to make significant progress toward shoring up our roadway and overall grading design throughout our residential subdivision design. We've expanded our understanding of how and why assemblies and subassemblies are considered to be the building blocks of all transportation-focused designs.

We also learned many additional settings and intricacies involved with furthering our roadway design by applying different site requirements and conditions so that our Corridor models can automatically make decisions within our design. We've also learned how we can merge Surface models from our land development and roadway modeling tool belts to create one comprehensive and fully integrated Surface model.

In our next chapter, we'll replace our roadway modeling tool belt with our utility modeling tool belt to finish up our residential subdivision design. As we understand utility modeling tools available to us in Civil 3D, we'll be able to design a storm drainage network, a sanitary sewer network, and a domestic water network that will not only support our new residential subdivision but also tie into the existing networks along York Highway.

12

Utility Modeling Tool Belt for Everyday Use

In the previous two chapters, we had an opportunity to unwrap all of the roadway modeling-focused tools available to us within Civil 3D, while also learning some very practical applications of how we can utilize them to further progress on our residential subdivision design. Not only did we learn how to use these particular tools to create design objects within Civil 3D, but we also continued to understand how best to manage these objects and the significance each one provides to the overall design.

In this chapter, we'll start off by combining and refining our proposed Surface model a little more by utilizing our surface analysis tools and utilize our land development tool belt to ensure that we have proper site drainage throughout our residential subdivision design. Once we are able to locate high and low elevations throughout our residential subdivision design, we'll then be able to properly locate our storm drainage inlets.

After our storm drainage design is complete, we can then jump into our sanitary sewer and domestic water main design that we'll tie into systems along York County Highway and will service our entire residential subdivision design. That said, in this chapter, we'll be covering the following topics:

- Refining proposed Surface models to accommodate proper site drainage
- Creating and modifying storm drainage pipe networks
- Creating and modifying sanitary sewer pipe networks
- Creating and modifying pressure networks

With that, let's go ahead and open up Civil 3D, or go to your start screen if already open, and create a new drawing using similar steps outlined in *Chapter 7, Alignments - The Second Foundational Component to Designs within Civil 3D*. We can use our `Company Template File.dwt` file located in `Practical Autodesk Civil 3D 2023\Chapter 12` and select **Open** in the lower right-hand corner of the **Select Template** dialog box. Once our new file is created, we'll want to save it as our `Utility Model.dwg` file to our `Practical Autodesk Civil 3D 2023\Chapter 12\Model` location.

As discussed back in *Chapter 3, Sharing Data within Civil 3D*, model files are intended to contain Civil 3D-modeled objects (both existing conditions and proposed design objects) that are broken out by the type of design objects contained within. Model files will also data reference design data from other model files and external reference files as overlays. Gravity and pressure-based utility network models are able to be data referenced across multiple files, and therefore should be created within our `Utility Model.dwg` file.

With that, let's go ahead and attach our `Survey Model.dwg` file as an overlay, contained within our `Practical Autodesk Civil 3D 2023\Chapter 12\Model` location, as well as our `Site Plan Reference.dwg` file as an overlay, contained within our `Practical Autodesk Civil 3D 2023\Chapter 12\Reference` location. Then, we'll want to jump back into our **Prospector** tab in our toolspace, and then set the **Working Folder** type of our **Data Shortcuts** project to the `Practical Autodesk Civil 3D 2023\Chapter 12` location and select the **C3D_2023_123456_Data_Shortcuts** project, as shown in *Figure 12.1*:

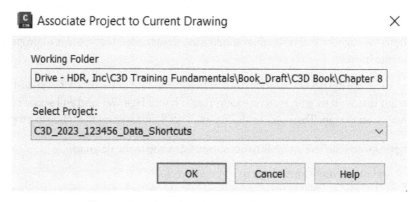

Figure 12.1 – Associate Project to Current Drawing

After our Civil 3D data shortcuts project has been associated with our current file, we can then safely create data references of **SRF - Proposed Grade - Residential Subdivision** and **SRF - Existing Grade – FromSurveyPoints** Surface models, along with our **Existing Sanitary Sewer**, **Existing Storm Drainage**, **Existing Water** gravity and pressure networks in our current `Utility Model.dwg` file.

To do this, we need to expand our **Alignments** category and our **Proposed Conditions** folder, right-click on each object (shown in *Figure 12.2*) and select the **Create Reference…** option for each:

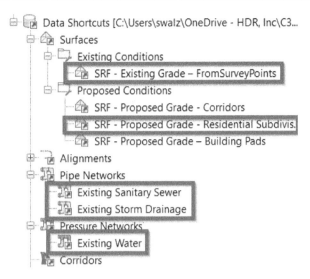

Figure 12.2 – Creating data references of identified Civil 3D objects

Then, with our `Utility Model.dwg` file set up for us to include all objects we have available to us to represent and reference the existing proposed built environment, we are now ready to continue designing our residential subdivision layout.

Technical requirements

The exercise files for this chapter are available at `https://packt.link/UoiPn`

Refining proposed Surface models to accommodate proper site drainage

Inside our `Utility Model.dwg` file, we'll want to create a new Surface model that will represent our complete proposed built environment. Although we have created a comprehensive **SRF - Proposed Grade - Residential Subdivision** Surface model that will represent our proposed portion of the built environment, we'll need to create a new surface that combines both existing and proposed conditions to more clearly understand how our entire residential subdivision design will drain during rain events.

That said, we'll use the following steps (similar to those defined in previous chapters) to create a full future-built Surface model that we can use to analyze and identify key locations where we'll need to place inlets and pipes:

1. Open up the toolspace from the **Home** ribbon.

2. Navigate to **Surfaces** within the **Prospector** tab.

3. Right-click and select the **Create Surfaces…** option.

4. In the **Create Surfaces** dialog box that appears, fill out the fields as follows:

    - **Type: TIN Surface**

    - **Name: SRF - Site Drainage - Residential Subdivision**

    - **Description: Combined Existing and Proposed Surfaces for Drainage Analysis**

    - **Surface Style: Elevation Banding (2D)**

    - **Render Material: ByBlock**

5. Once created, we'll use steps detailed in earlier chapters to paste our existing and then proposed surfaces to the surface definition via **Edits**.

6. Right-click on the **SRF - Site Drainage - Residential Subdivision** Surface model in **Prospector** and select the **Edit Surface Style…** option.

7. Navigate to the **Analysis** tab.

8. Expand **Elevations**.

9. Change the **Range Color Scheme** type to **Hydro**.

10. Click **OK**.

The final display should look similar to that shown in *Figure 12.3*. Based on the thematic coloring applied to the default elevation ranges, we know that darker blue areas represent higher elevations within the Surface model, whereas light blue represents lower elevations:

Figure 12.3 – Elevation banding display for SRF - Site Drainage - Residential Subdivision Surface model

With our elevation banding displayed, let's take a look at our existing storm drainage gravity pipe network. As we select our individual structures, right-click and select **Structure Properties…**, and we'll gain access to our **Structure Properties** dialog box, at which point we'll want to take note of the invert elevations of the connected pipes to each structure within the **Connections** tab (refer to *Figure 12.4*):

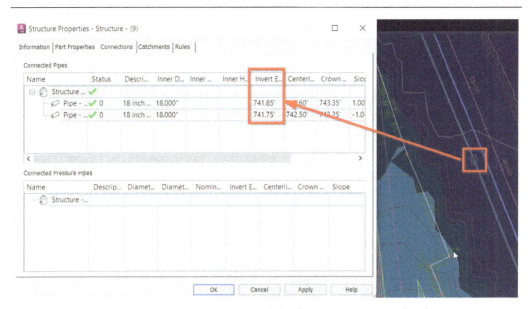

Figure 12.4 – Structure Properties dialog box pipe connection details

After checking the inverts of connected pipes of several of our structures, we can quickly determine that the flow of the entire existing storm drainage gravity network is flowing from south to north along York Highway, with the lowest and most logical structure to tie in being at **Structure - (10)**, which contains an **Invert Out Pipe** connection at an elevation of 739.65', as shown in *Figure 12.5*:

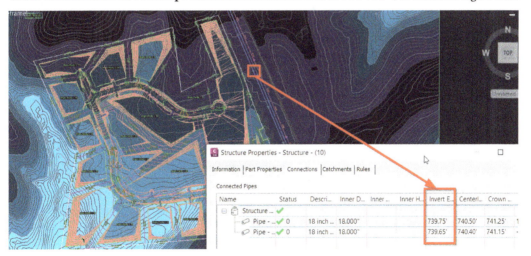

Figure 12.5 – Lowest invert out at Structure - (10)

Taking note of the invert-out elevation of 739.65', we can make an attempt to drain our site above 741.00' (adding almost 1.5') to ensure that we have enough slope for stormwater to drain through our proposed storm drainage gravity network where possible.

The remainder of our site that falls below the 741.00' elevation will need to drain to a detention pond that will hold water during rain events to contain and then slowly release downstream in an effort to decrease erosion and improve water quality (refer to local guidelines and requirements while designing these).

With that, let's go back to our **SRF - Site Drainage - Residential Subdivision** Surface model and adjust the thematic elevation mapping that is currently being displayed so that we can clearly understand where the 741.00' delineation runs through our site.

Let's select our **SRF - Site Drainage - Residential Subdivision** Surface model (either in **Model Space** or in our **Prospector**), right-click on our mouse, and go to the **Surface Properties...** option. Once our **Surface Properties** dialog box appears, we'll jump over to our **Analysis** tab and make the following selections (also shown in *Figure 12.6*), and then click on the **OK** button:

- **Analysis type**: **Elevations**
- **Number of ranges**: 2
- ID 1 **Minimum Elevation**: 671.78'
- ID 1 **Maximum Elevation**: 741.00'
- ID 2 **Minimum Elevation**: 741.00'
- ID 2 **Maximum Elevation**: 794.00'

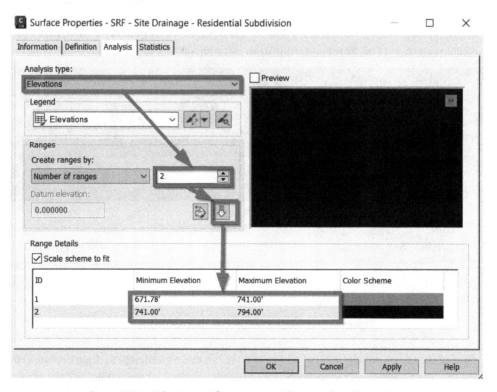

Figure 12.6 – Adjusting surface property elevation banding values

With our updated thematic mapping displaying, let's make two more adjustments to the **SRF - Site Drainage - Residential Subdivision** Surface model so that we can quickly determine the surface areas within each of the elevation ranges that we'll need to drain during storm events.

To do so, let's create a quick polyline boundary line around the entire extent of the residential subdivision and then apply an outer boundary to our **SRF - Site Drainage - Residential Subdivision** Surface model, as shown in *Figure 12.7*:

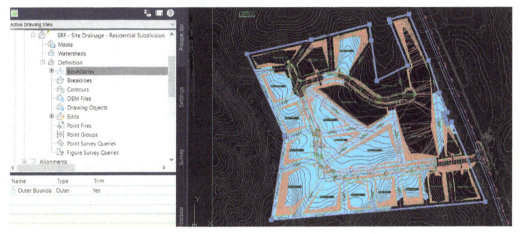

Figure 12.7 – Applying boundary to SRF - Site Drainage - Residential Subdivision Surface model

Next, we'll select our **SRF - Site Drainage - Residential Subdivision** Surface model, right-click with our mouse, and select the **Surface Properties…** option. Jumping back over to the **Analysis** tab, we'll make the following adjustments (also displayed in *Figure 12.8*):

- **Analysis type**: **User-defined contours**
- **Ranges | Number**: 1
- ID 1 **Elevation**: 741.00'

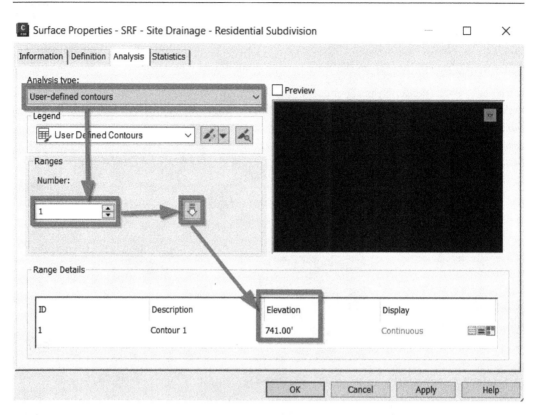

Figure 12.8 – Applying user-defined contour analysis to surface

Then, we'll hop over to the **Information** tab and make the following selections (as shown in *Figure 12.9*):

1. Switch to the **Information** tab within the **Surface Properties** dialog box.
2. Create a new surface style, if one doesn't already exist, and call it `User-Contour Only`.
3. In the **Edit Surface Style** dialog box, we'll switch over to the **Display** tab.
4. Turn on only the **User Contours** option in the **Plan View Direction**.
5. Click **OK**, with the final display appearing similar to that shown in *Figure 12.9*:

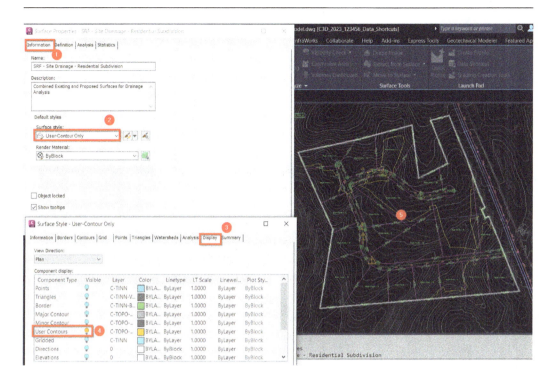

Figure 12.9 – Applying user-defined contour display style to surface

Now, we are able to use the **Extract Objects from Surface** tool to extract the user-defined contours from the surface.

We'll then use the extracted user-defined contours, our residential subdivision site boundary, along with new polylines that will need to be created using basic AutoCAD tools to determine pervious versus impervious areas versus disturbed areas versus total areas throughout our site to determine catchment areas, inlet locations, and the best route for our proposed storm drainage design.

This combination of linework can be utilized later on by the design engineer to further analyze surface conditions for storm drainage purposes. Without getting into the weeds on the engineering aspects, let's focus on the actual gravity network design itself, supporting both storm drainage and sanitary sewer systems.

In the next section, we'll do just that and begin laying out our design networks accordingly.

Creating and modifying storm drainage pipe networks

After running through several calculations and ensuring that we're adhering to local and state design requirements and regulations, we've been able to determine that we'll need a detention pond with an approximate storage of 22,700 cubic feet, and it will need to be located near the lowest point of our site, which ends up being in the southwestern-most lot within our residential subdivision design, as shown in *Figure 12.10*:

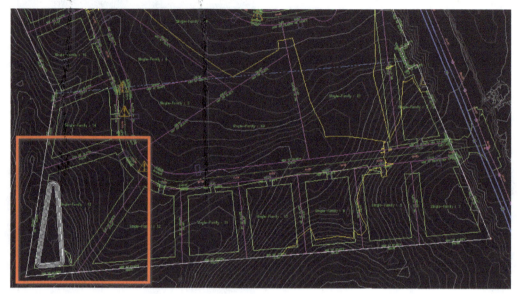

Figure 12.10 – Location of the detention pond

Let's go ahead and jump into our `Grading Model.dwg` file to include this new Detention Pond as a standalone surface and then paste it into our comprehensive **SRF - Proposed Grade - Residential Subdivision** Surface model. Using our land development tool belt, we'll want to create a new site for this detention pond.

Feel free to create a boundary polyline that delineates the top of the pond and convert it to a feature line using the **Create Feature Lines from Object** tool we reviewed in *Chapter 9, Land Development Tool Belt for Everyday Use*. We can achieve this with the accompanying **Feature Lines**, **Grading Group**, and **Grading** objects, for ultimate inclusion into our overall **SRF - Proposed Grade - Residential Subdivision** Surface model.

Please keep in mind that this process can be quite iterative to ensure that we are creating a pond that will allow us to store approximately 22,700 cubic feet per design requirements.

With these elements, we can determine invert elevations of pipes associated with our proposed storm drainage gravity network, with the final grading model potentially looking similar to that shown in *Figure 12.11* (depending on the **Top of Pond** feature line you started with):

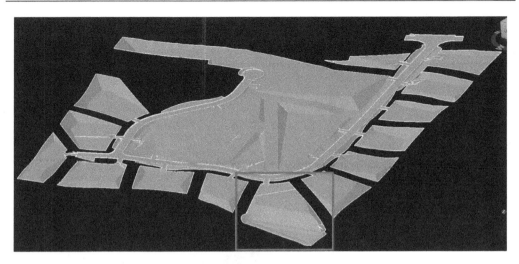

Figure 12.11 – Graded detention pond

Jumping back into our jump into our `Utility Model.dwg` file, go ahead and sync your data shortcuts to ensure that we are referencing the most recent **SRF - Proposed Grade - Residential Subdivision** Surface model so that we can begin laying out our proposed storm drainage network.

With that, let's go ahead and begin familiarizing ourselves with the gravity network design tools that Civil 3D has available. Fortunately for us, the Civil 3D template that we've been using already includes a lot of the standard pipes and structures available out of the box. We'll begin using the out-of-the-box parts available and review how to create new customized parts lists in the final section of this chapter.

Jumping up to our **Home** ribbon and navigating over to the **Create Design** panel, we'll notice a down arrow next to **Pipe Network**. If we select the down arrow, we have several ways we can generate both gravity and pressurized utility networks available to us, as displayed in *Figure 12.12*.

Running from top to bottom, we have the following tools:

- **Pipe Network Creation Tools**: Allows us to manually create gravity utility networks throughout our site

- **Create Pipe Network from Object**: Allows us to convert AutoCAD lines, arcs, polylines, and feature lines to a gravity utility network

- **Pressure Network Creation Tools**: Allows us to manually create pressurized utility pipe networks throughout our site

- **Create Pressure Network from Object**: Allows us to convert AutoCAD lines, arcs, polylines, and feature lines to a pressurized utility network

- **Create Pressure Network from Industry Model**: Allows us to convert water industry models to a pressurized utility network:

Figure 12.12 – Pipe Network tools

With that, let's go ahead and activate the **Pipe Network Creation Tools** option to pull up our **Create Pipe Network** dialog box. In the **Create Pipe Network** dialog box, we'll go ahead and fill out the fields and make selections as follows (also shown in *Figure 12.13*), and then select the **OK** button at the bottom of the dialog box:

- **Network name: GPN - Proposed Storm Drainage - York Hwy**

- **Network description: Proposed Storm Drainage Network that will tie-into York Hwy Existing Drainage Network and our Proposed Detention Pond**

- **Network parts list: Storm Sewer**

- **Surface name: SRF - Proposed Grade - Residential Subdivision**

- **Structure label style: Data with Connected Pipes (Storm)**

- **Pipe label style: Length Description and Slope**

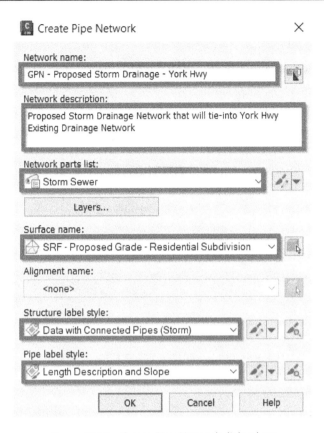

Figure 12.13 – Create Pipe Network dialog box

After selecting the **OK** button in the **Create Pipe Network** dialog box, we'll be presented with our **Network Layout Tools** toolbar, which will provide us with access to many of the necessary gravity utility network design tools that Civil 3D has available. Running from left to right, we have the following tools available to us (also shown in *Figure 12.14*):

1. **Pipe Network Properties**: Allows us to access the **Pipe Network Properties** dialog box where we can specify various settings and styles to be applied to our gravity utility network being designed.

2. **Select Surface**: Allows us to specify a surface that we intend to reference and target as we lay out our proposed gravity utility network.

3. **Select Alignment**: Allows us to specify an alignment already created, or available, in our current file. Alignment referencing can help with staking purposes where we can quickly identify the station and offset positioning.

4. **Parts List**: Allows us to specify a parts list that contains a predefined set of structures and pipes.

5. **Structure List**: Allows us to specify the type of structure we'd like to place in our files as we lay out our gravity utility network.

6. **Pipe List**: Allows us to specify the type of pipe we'd like to place in our files as we lay out our gravity utility network.

7. **Structure Connection and Pipe Insertion Points**: Allows us to change the placement and insertion points of structures and connected points as we lay out our gravity utility network.

8. **Toggle Upslope/Downslope**: Allows us to change the flow direction of our gravity utility network.

9. **Delete Pipe Network Object**: Allows us to remove a structure and/or pipe from our gravity utility network being designed.

10. **Draw Pipes and Structures**: Allows us to design our gravity utility network as pipes only, structures only, or pipes and structures all at once.

11. **Pipe Network Vistas**: Allows us to pull up the **Panorama** where we can make modifications to structures and/or pipes of the gravity utility network being designed.

12. **Undo**: Allows us to undo previous insertion points of structures and/or pipes:

Figure 12.14 – Create Pipe Network dialog box

> **Note**
>
> The **Network Layout Tools** toolbar has a tendency to close out if users accidentally hit the *Esc*, spacebar, and/or *Enter* keys on the keyboard. If that does occur, we can always recall the **Network Layout Tools** toolbar by right-clicking with our mouse on the **Network Name** field in our **Toolspace** | **Prospector** and selecting **Edit Network**.

For our **GPN - Proposed Storm Drainage** gravity utility network, we'll start at the northern side of our residential subdivision design to place inlets and pipes at key locations to divert stormwater into the **Existing Storm Drainage** gravity utility network along York Hwy.

As we analyze our future graded conditions, we can see where we have ditches alongside our road (beyond the paved surface) in between the curb and gutter and our building pads, and quickly determine where our stormwater is running to.

This will allow us to place our structures and connected pipes along with the placement of our driveway culvert pipes, and also identify locations where we will need to capture additional runoff via diversion ditches/swales.

If at any point you are having some difficulty trying to understand how water is flowing on top of our future graded conditions, Civil 3D has a pretty handy tool called **Water Drop**. The **Water Drop** tool will allow us to quickly analyze surface conditions and then draw a construction line indicating the most direct path that stormwater will eventually drain to based on the location we click within our residential subdivision design.

To access the **Water Drop** command, we can simply select our **SRF - Site Drainage - Residential Subdivision** Surface model to access our Surface Contextual ribbon along the top of our Civil 3D session. Once visible, navigate to our **Analyze** ribbon and select the **Water Drop** tool. Once activated, a **Water Drop** dialog box will appear, where we'll want to make the following selections (also displayed in *Figure 12.15*):

- **Path Layer: C-HYDR-CTCH-FPTH**
- **Path Object Type: 2D Polyline**
- **Place Marker at Starting Point: Yes**
- **Start Point Marker Style: Basic**

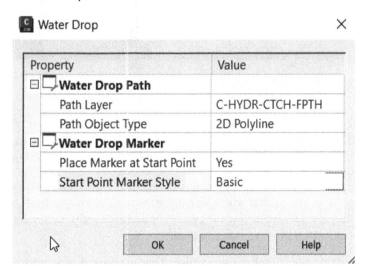

Figure 12.15 – Water Drop dialog box

> **Tip**
> If 3D polylines were selected, potentially leverage them to define our time of concentration for stormwater analysis purposes.

After setting these as our **Water Drop** defaults and clicking **OK** to exit out of the **Water Drop** dialog box, we'll be prompted at our command line to select a point.

As we select points throughout our residential subdivision design, Civil 3D is analyzing Surface model conditions, determining how stormwater will travel, and identifying where it will end up. An example of the output will look similar to that displayed in *Figure 12.16*:

Figure 12.16 – Water Drop analysis

After completing the northern portion of our **GPN - Proposed Storm Drainage** gravity utility network that ties into the **Existing Storm Drainage** gravity utility network along York Hwy, we'll run through a similar process to drain the remainder of our site to the proposed detention pond in the southwest corner that we designed earlier in this chapter.

You may have noticed that in locations where we are placing our driveway culverts, which are only displayed as pipes, the pipe rules are forcing the pipes to be designed with a minimum of 3' coverage, and not locating them with an invert elevation at grade.

Applying pipe rules within Civil 3D allows us to validate our gravity utility network designs as we're laying them out in Model Space. We can specify and set various design parameters that our networks will need to conform to, giving us a higher level of assurance that we are designing in accordance with codes and regulations.

Typically, when I'm laying out my proposed storm drainage designs where driveway culverts are required to properly drain the site, I'll place them temporarily using the same rules as a typical storm drainage network would be expected to conform to. Once all driveway culverts have been placed in their appropriate locations, I'll then create a new pipe rule that I will apply to my driveway culverts only.

To do this, we'll jump over to our **Settings** tab within our **Prospector**, navigate down to—and expand—our **Pipe** category to access our **Pipe Rule Sets** section. Once there, we'll right-click with our mouse on **Pipe Rule Set** and select the **New...** option, as shown in *Figure 12.17*:

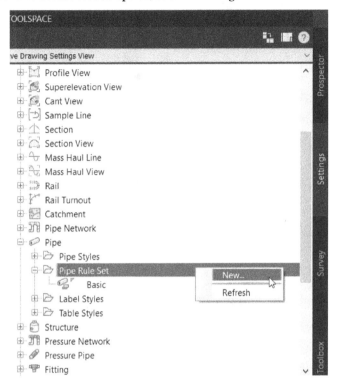

Figure 12.17 – Creating a new pipe ruleset

In the **Information** tab within the **Pipe Rule Set** dialog box that appears, we'll fill in the **Name** value as `Driveway Culverts` and **Description** value as **To be applied to Driveway Culvert Pipes with no Structure Connections**. Next, we'll switch over to our **Rules** tab and use the following steps (shown in *Figure 12.18*) to create a driveway culvert rule that we'll use to update our driveway culverts so that the inverts are at grade:

1. Select the **Add Rule...** option within the **Pipe Rule Set** dialog box.

2. Change the **Rule name** value to `Cover Only` in the **Add Rule** dialog box.

3. Click **OK** in the **Add Rule** dialog box.

4. Expand the **Cover Only** parameter in the **Pipe Rule Set** dialog box.

5. Change the value for **Minimum Cover** and **Maximum Cover** to `-1.50'` within the **Pipe Rule Set** dialog box, where the `1.50'` value actually places the driveway culvert pipes such that `18"` diameter pipe inverts are at grade.

6.   Click **OK** in the **Pipe Rule Set** dialog box:

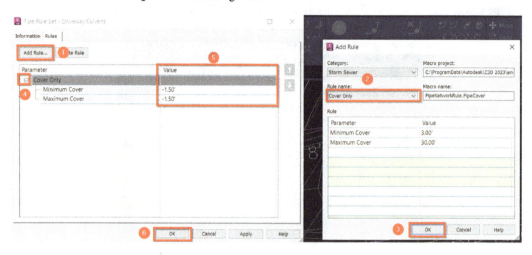

Figure 12.18 – Steps to creating a new pipe rule

With our new driveway culverts pipe rule created, we can go back to our storm drainage design and update the driveway culverts within our **GPN - Proposed Storm Drainage** gravity utility network.

To update these, we'll want to select all driveway culvert pipes displayed in our Model Space, then go back up to our **Pipe Network Contextual** ribbon and select the **Split Network** option within the **Modify** panel, as shown in *Figure 12.19*:

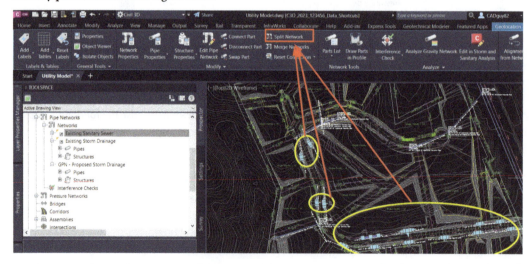

Figure 12.19 – Split Network tool

Once the **Split Network** tool has been initiated, we'll be prompted at the command line to either create a new pipe network or select a network. Since we haven't created a new pipe network to associate our driveway culverts with, we'll want to select the **Create new Pipe Network** option, which will then pull up our **Create Pipe Network** dialog box.

Here, we'll fill out the available fields and make selections as follows (also displayed in *Figure 12.20*), and then select **OK**:

- **Network name: GPN - Proposed Driveway Culverts**
- **Network description: Proposed Storm Drainage Network for Driveway Culverts Only**
- **Network parts list: Storm Sewer**
- **Surface name: SRF - Site Drainage - Residential Subdivision**
- **Structure label style: <none>**
- **Pipe label style: Length Description and Slope**

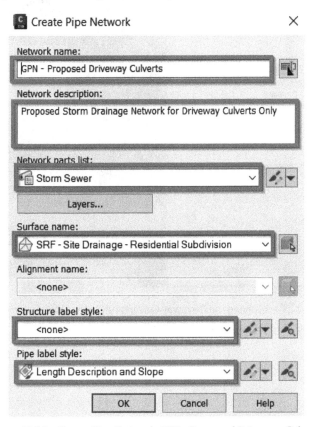

Figure 12.20 – Create Pipe Network: GPN - Proposed Driveway Culverts

With our driveway culverts now separated into their own gravity utility network, we can quickly assign and then apply our new driveway culverts pipe rule to all pipes to update cover and inverts accordingly.

Jumping back to our **Prospector** tab in our toolspace, let's scroll down to—and expand—our **Pipe Networks** category, expand our **GPN - Proposed Driveway Culverts** gravity utility network, and then select the **Pipes** category within.

In the lower portion of our toolspace, we'll want to select all pipes listed, then, while still holding the *Shift* key on our keyboard, we'll click on one of the fields within the **Rule Set** column where it says **Basic**.

By selecting the field, we can change our **Basic** ruleset to the newly created **Driveway Culverts** ruleset and click **OK** to apply to all driveway culvert pipes within our **GPN - Proposed Driveway Culverts** gravity utility network, as shown in *Figure 12.21*:

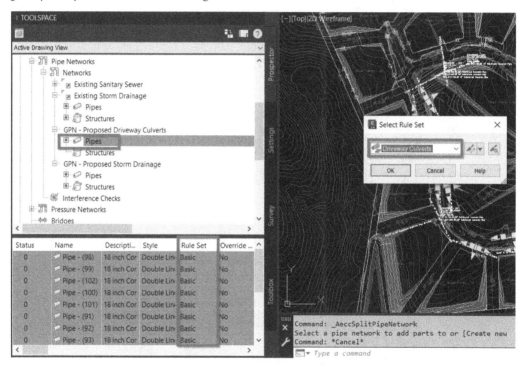

Figure 12.21 – Applying the Driveway Culverts ruleset to pipes within the
GPN - Proposed Driveway Culverts gravity utility network

With our driveway culvert pipes now reading the new **Driveway Culverts** ruleset, we can now individually select each pipe within our Model Space to call up our **Pipe Network Contextual** ribbon again, then select the **Modify** down arrow in the **Modify** panel, and select **Apply Rules** (refer to *Figure 12.22*):

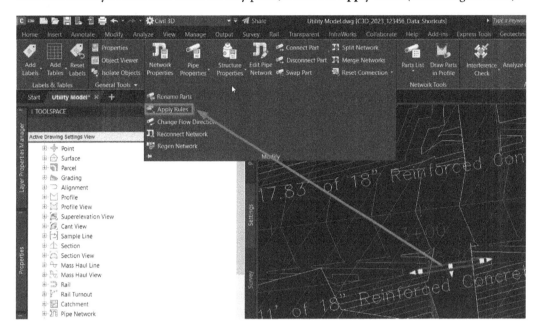

Figure 12.22 – Applying rules to pipes within the GPN - Proposed Driveway Culverts gravity utility network

At this point, we should be in pretty good shape with our storm drainage gravity utility network design within our residential subdivision. Now, let's focus on our proposed sanitary sewer network that we'll tie into our existing sanitary sewer gravity network located along York Hwy.

Creating and modifying sanitary sewer pipe networks

As recognized while analyzing elevations and grades at the beginning of our storm drainage gravity utility network design, we realized that we are approximately 25' below the existing grade along York County Highway at our lowest point within our residential subdivision design.

That said, we will need to use a combination of gravity and pressurized utility networks and find a suitable location for the placement of a required pump station to ensure that we are able to remove wastewater from our site and tie into the existing sanitary sewer network along York Hwy.

With that, let's go ahead and create a proposed sanitary sewer network using similar steps used to create our **GPN - Proposed Storm Drainage** gravity utility network earlier in this chapter.

Using the **Pipe Network Creation Tools** option as done earlier, we'll create a new sanitary sewer gravity utility network with the following fields applied (also displayed in *Figure 12.23*) and then click the **OK** button:

- **Network name**: **GPN - Proposed Sanitary Sewer**
- **Network description**: **Proposed Sanitary Sewer Network that will tie-into Proposed Pump Station**
- **Network parts list**: **Sanitary Sewer**
- **Surface name**: **SRF - Proposed Grade - Residential Subdivision**
- **Structure label style**: **Data with Connected Pipes (Sanitary)**
- **Pipe label style**: **Length Description and Slope**

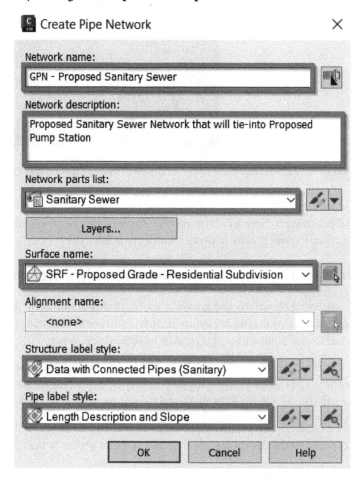

Figure 12.23 – Create Pipe Network dialog box

Starting in the cul-de-sac, we'll begin laying out our **GPN - Proposed Sanitary Sewer** gravity utility network west through the intersection of our **ALG – Subdivision Side Road – Cul-De-Sac** and **ALG – Subdivision Main Road – Access** alignments, and then south toward our southwestern-most lot where we'll end up placing our pump station.

Once the northern portion of our **GPN - Proposed Sanitary Sewer** gravity utility network is completely laid out, we'll then want to work our way from the first lot near the main intersection with York Hwy to the same southwestern-most lot, with the final **GPN - Proposed Sanitary Sewer** gravity utility network looking similar to that shown in *Figure 12.24*:

Figure 12.24 – GPN - Proposed Sanitary Sewer gravity utility network

Before moving on to designing a pressurize sanitary sewer force main that will essentially carry wastewater from our pump station into the **Existing Sanitary Sewer** gravity utility network within York Hwy, we'll want to perform an interference check between our storm drainage and sanitary sewer gravity utility networks to ensure that we have enough clearance between utility networks.

To access this, we'll want to go up to our **Analyze** ribbon and select the **Interference Check** tool located in the **Design** panel. Once the **Interference Check** tool has been activated, we'll see a **Create Interference Check** dialog box appear, at which point we can fill out the fields as necessary and define the interference checking criteria by selecting the **3D proximity check criteria…** option, as shown in *Figure 12.25*:

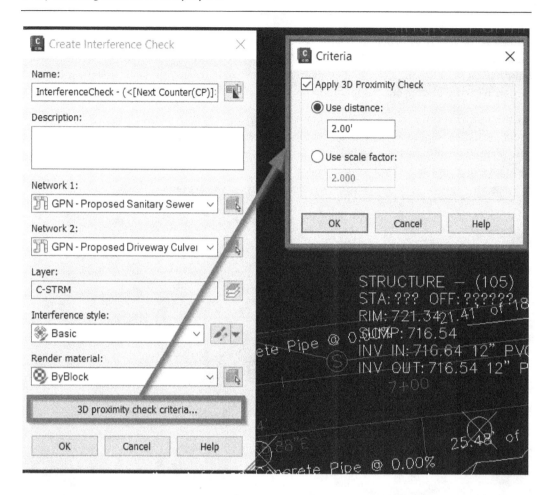

Figure 12.25 – Create Interference Check dialog box

After performing, we will receive a message that will notify us whether we have any interferences. If interferences have been detected, a node, or block, will be placed at the exact location in which there is a conflict that does not meet the **3D Proximity Check Criteria** values we had input originally.

If any interferences have been detected, it's recommended to resolve these by selecting **Connected Structures**, right-clicking with our mouse in Model Space, selecting the **Structure Properties** option, and updating the **Connected Pipe** inverts accordingly.

After all detected interferences have been resolved, we can focus on designing our pressurized systems for our required sanitary sewer force main along with our domestic water main to service all individual properties.

Creating and modifying pressure networks

With the gravity utility network portion of our sanitary sewer designed, let's begin laying our force main that will carry wastewater from our pump house located in the southwestern-most lot and tie to the **Existing Sanitary Sewer** main located along York Hwy.

As discussed earlier in the chapter, we'll want to jump back up to our **Create Design** panel within our **Home** ribbon and activate the **Pressure Network Creation** tools (refer to *Figure 12.12*).

Once the **Pressure Network Creation** tools have been activated, we'll be presented with a **Create Pressure Pipe Network** dialog box where we'll want to fill out the fields and make the selections as follows (also shown in *Figure 12.26*), and then click the **OK** button:

- **Network name: PPN - Proposed Sanitary Sewer**
- **Network description: Proposed Force Main from Pump Station within Residential Subdivision to tie into Existing Sanitary Sewer Gravity Utility Network along York Hwy**
- **Pipe Run name: Force Main**
- **Parts list: Water**
- **Pipe size: pipe-6 in-push on-ductile iron 350 psi-AWWA C150**
- **Reference surface: SRF - Proposed Grade - Residential Subdivision**
- **Create surface profile to follow:** *Check*
- **Cover:** 3.00'
- **Pressure pipe label style: Nominal Diameter and Material**
- **Fitting label style: Nominal Diameter Bend Angle and Material**
- **Appurtenance label style: Nominal Diameter Valve Type**

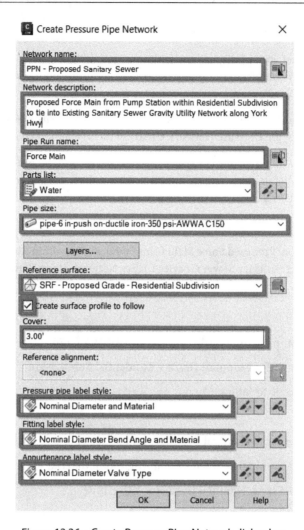

Figure 12.26 – Create Pressure Pipe Network dialog box

After all fields and selections have been made and we click the **OK** button inside of the **Create Pressure Pipe Network** dialog box, we'll want to bring our attention to the **Pressure Pipe Network Contextual** ribbon that will automatically display along the top of our Civil 3D session, as shown in *Figure 12.27*. Running from left to right, we have the following panels and tools available to us:

- **Properties**:

  - **Network Properties**: Allows us to modify general, layout, profile, and section properties associated with our pressure pipe network

  - **Part Properties**: Allows us to edit individual parts associated with our pressure pipe network

- **Pipe Run**:

  - **Add New Pipe Run**: Allows us to create a new pipe run as needed. This tool will be required to begin laying out our design and will also come in handy if there is a need to branch and separate our full pressure pipe network into individual runs later on

  - **Pressure Pipe Selection**: Allows us to select the parts that we would like to apply to our design

  - **Add Bends Automatically**: Allows us to either automatically add bend fittings or have pipes connect directly to each other

- **Layout**:

  - **Add Bend/PI**: Allows us to insert a bend and/or point of intersection into our design and alignment as needed

  - **Remove Bend/PI**: Allows us to remove a bend and/or point of intersection from our design and alignment as needed

  - **Add Branch Fitting**: Allows us to insert a branch fitting in the event that we need to create a new pipe run

  - **Add Fitting**: Allows us to insert fittings individually based on the part selection directly across from this tool

  - **Add Appurtenance**: Allows us to insert an appurtenance individually based on the part selection directly across from this tool

  - **Panorama**: Allows us to pull up our **Panorama** to edit pipes, fittings, and appurtenance properties

- **Profile**:

  - **Pipe Run Profile**: Allows us to create and modify a profile along our pipe run

  - **Draw Parts in Profile View**: Allows us to project parts within our network into a profile view already created within our drawing

- **Modify**:

  - **Swap Parts**: Allows us to swap parts in the event that we need to change diameter, material, and so on

  - **Break Pipe Run**: Allows us to segment a pipe run as needed

  - **Merge Pipe Runs**: Allows us to merge a pipe run as needed

  - **Relink Pipe Run**: Allows us to relink our pipe run with its corresponding alignment in the event that it has been broken

  - **Regen Pressure Solids**: Regenerates solids displayed for fittings and appurtenances in the event that they are missing

- **Connect to Structure**: Allows us to connect a pressure pipe to a gravity structure (this will be needed for our force main connection to the **GPN - Proposed Sanitary Sewer** gravity utility network)

- **Analyze**:

  - **Design Check**: Allows us to validate our design is in conformance with the defined parameters and tolerances specified

  - **Depth Check**: Allows us to verify that our pressure pipe network meets the minimum depth defined

- **Labels & Tables**:

  - **Add Labels**: Allows us to add labels associated with our pressure pipe network design

  - **Add Tables**: Allows us to create tables displaying information associated with our pressure pipe network design

- **Compass**:

  - **Visibility**: Allows us to toggle on/off our design compass that will be displayed as we are laying out our pressure pipe network design

  - **Snapping**: Allows us to toggle on/off the ability to snap to specified bend/fitting angles as we are laying out our pressure pipe network design

- **Close**: Allows us to close out of the **Contextual** ribbon:

Figure 12.27 – Pressure Pipe Network contextual ribbon

> **Note**
> After initially creating our pressure pipe network, we will be prompted at the command line to specify the first pressure pipe point to begin laying out our design. If you may have accidentally canceled out of this prompt, we'll need to select the **Add New Pipe Run** tool within our **Pipe Run** panel. Once activated, a **Create Pipe Run** dialog box will appear, at which point we'll want to fill out fields and make selections similar to those detailed in *Figure 12.26*.

Let's go ahead and start laying out our force main pipe run by snapping to the center of our final structure created inside of our pump station for our **GPN - Proposed Sanitary Sewer** gravity utility network. After snapping to the center and clicking with the left button on our mouse to begin our layout, we'll run perpendicularly into our wet well right beside the pump station. Next, we'll then run 90 degrees toward our **ALG – Subdivision Main Road – Access** alignment and run our force main pipe along the south side of our subdivision main road and tie into the existing sanitary sewer structure at the main entrance of our subdivision.

After our network has been configured, we'll hop back up to our **Pressure Pipe Network Contextual** ribbon and select the **Connect to Structure** tool located within our **Modify** ribbon pulldown. Once the **Connect to Structure** tool has been activated, we'll be prompted at our command line to select a structure from our gravity utility network and a pipe from our pressure utility network to connect to.

We're going to then make connections at the beginning and end of our overall force main pipe run accordingly, with the final force main design looking similar to that shown in *Figure 12.28*:

Figure 12.28 – Force main pipe run design

Next, we'll move into designing our domestic water main. Taking local design requirements into consideration, there will likely be a minimum horizontal and vertical separation that needs to be maintained between other utilities, with an emphasis on sanitary sewer networks.

That said, we'll need to snake our domestic water main alongside our roads through our residential subdivision design with the tie back into the existing water main along York Hwy.

So, let's jump back up to our **Create Design** panel within our **Home** ribbon and select the **Pressure Network Creation** tools again and fill out the new **Pressure Network** dialog box as follows (also displayed in *Figure 12.29*), and then select the **OK** button:

- **Network name: PPN – Domestic Water Main**
- **Network description: Proposed Domestic Water Main within Residential Subdivision to tie into Existing Water Pressure Utility Network along York Hwy**
- **Pipe Run name: Water Main**
- **Parts list: Water**

- **Pipe size: pipe-6 in-push on-ductile iron 350 psi-AWWA C150**
- **Reference surface: SRF - Proposed Grade - Residential Subdivision**
- **Create surface profile to follow:** *Check*
- **Cover:** 3.00'
- **Pressure pipe label style: Nominal Diameter and Material**
- **Fitting label style: Nominal Diameter Bend Angle and Material**
- **Appurtenance label style: Nominal Diameter Valve Type**

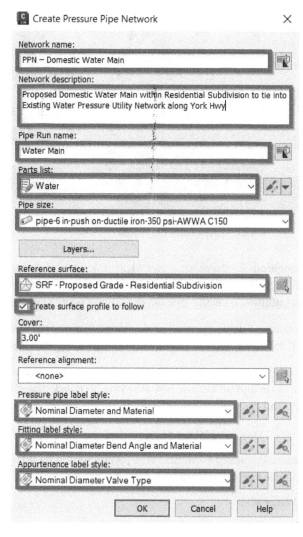

Figure 12.29 – Create Pressure Pipe Network dialog box

After selecting the **OK** button, let's go ahead and start laying out our **PPN – Domestic Water Main** pressure utility network, starting at the tie-in connection point of our **Existing Water** pressure utility network along York Hwy.

We'll start by snapping to a point along the centerline of the **Existing Water** pressure utility network along York Hwy and work our way through our residential subdivision to the cul-de-sac. After completion, we'll zoom back to our connection point to the **Existing Water** pressure utility network along York Hwy to add a connecting tee.

To do this, let's go back up to our **Pressure Network Contextual** ribbon, change our **Fitting Selection** value to **tee-12 in x 6 in- push on-ductile iron–350 psi-AWWA C111/C153**, and select the **Add Fitting** option within our **Layout** panel, as shown in *Figure 12.30*:

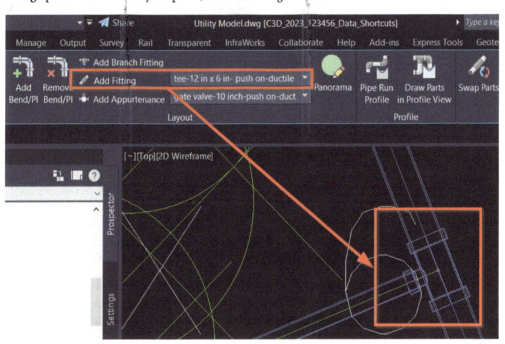

Figure 12.30 – Adding fitting only to the pressure network

After finalizing this connection point, we should end up with a domestic water main design looking similar to that displayed in *Figure 12.31*:

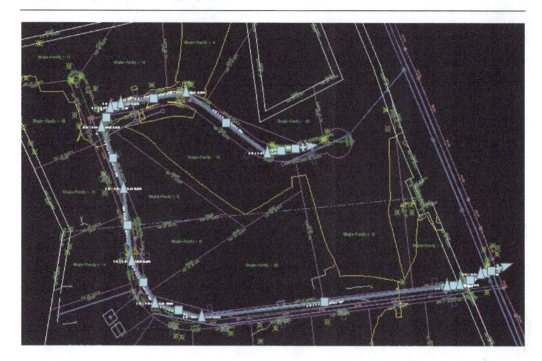

Figure 12.31 – Final display of our domestic water main pressure network

At this point, if we were to select our pressure networks and our surface to view in **Object Viewer**, we'd quickly notice that although the endpoints of our pipes, along with our fittings, are located underneath our surface, there are still many locations along the length of our pipes that are above our surface. To update, we'll want to leverage our profile views to segment these long runs and push our pipes below the surface and ensure that we have enough coverage.

To do this, we'll start by selecting our **PPN – Domestic Water Main** pressure pipe network, either in Model Space or in Toolspace, and select **Edit Pressure Network**. With our **Pressure Pipe Network** contextual ribbon activated, let's select the **Pipe Run Profile** tool within our **Profile** panel, as shown in *Figure 12.32*:

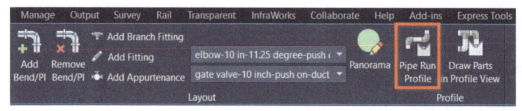

Figure 12.32 – Pipe Run Profile tool

Once the **Pipe Run Profile** tool has been activated, a **Pipe Run Profile Settings** dialog box will appear. In this dialog box, we'll want to fill out fields and make selections as follows (also displayed in *Figure 12.33*), and then click the *Enter* button at the bottom of the dialog box:

- **Offset Style**: **Cut Length**

- **Reference Profile**: **SRF - Proposed Grade - Residential Subdivision Surface**

- **Offset Distance**: 3.00'

- **Update Dynamically**: *Check the box*

- **Draw Profile in**: **New Profile View**

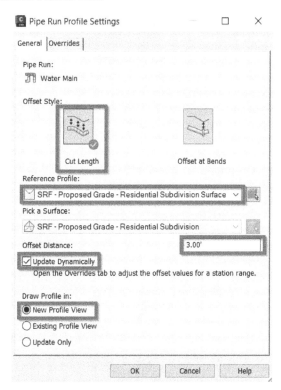

Figure 12.33 – Pipe Run Profile Settings dialog box

After selecting **OK**, a **Create Profile View** dialog box will appear. Since this is only a temporary profile view that we're creating at this point in time to simply segment our pressure pipes and push down to a consistent depth below our **SRF - Proposed Grade - Residential Subdivision** Surface model, we'll simply just select the **Create Profile View** option along the bottom of our **Create Profile View** dialog box without filling out any additional criteria. We'll then be prompted to select **Profile View Origin** at the command line. Once our origin has been defined, we should have a profile view looking similar to that shown in *Figure 12.34*:

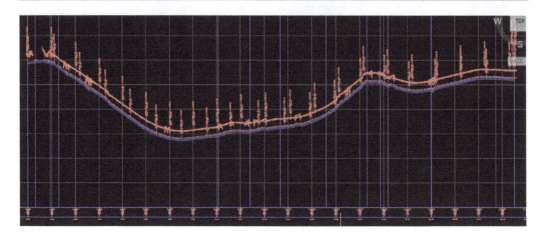

Figure 12.34 – Pipe Run profile view

As shown in *Figure 12.34*, we have successfully segmented our pipes and now have a consistent 3.00' coverage applied throughout the entire run of our **PPN - Domestic Water Main** pressure pipe network. Using the same steps, we want to run through the process for our **PPN - Proposed Sanitary Sewer** pressure pipe network as well.

With the final pieces to our residential subdivision design in place, we can begin exploring different ways we can analyze our design, with the end goal of producing a comprehensive plan set that we are confident conforms to all design standards and requirements necessary.

Summary

As we worked through this chapter, we have been able to make significant progress toward shoring up our utility design throughout our residential subdivision. We've learned the many aspects of utility design as it relates to both gravity and pressure utility networks.

We were able to perform several modifications and checked our designs against various design requirements and parameters. We even learned how to tie our proposed utility networks into existing networks, as well as tie gravity and pressure utility networks into each other.

In our next chapter, we'll take a look at sectioning capabilities and tools available to us within Civil 3D, where we can further analyze the residential subdivision grading, roadway, and utility designs that we've developed thus far. This will be the final form of analysis that we'll be covering before jumping into plan production, sheet creation, and extended capabilities available to us to increase efficiencies along the way.

Part 4:
Advanced Capabilities with
Civil 3D

In this next part, we will dive deeper into some of the advanced tools and functionality built into Civil 3D that will streamline your final design efforts and provide a new level of consistency and efficiency throughout the entire project design life cycle.

The following chapters are included in this section:

- *Chapter 13, Section Creation and Analysis*
- *Chapter 14, Automating Sheet Creation*

Section Creation and Analysis

Up to this point, we've essentially been chipping away at progressing our Residential Subdivision design, all while learning many of the design and analysis tools available to us within Civil 3D. We've taken an existing site composed of two existing lots and have subdivided it into smaller building parcels, graded lots and building pads, two roads, and several utilities to service future development. From a design standpoint, we are just about complete. From an analysis standpoint, we'll take a look at one more set of tools, Sectioning tools, available to us within Civil 3D that will allow us to further analyze our design.

Also, up to this point, we've mostly been working in plan view with the occasional need to view our modeled objects in our Object Viewer. As we utilize the Sectioning tools, we'll be able to view our modeled objects through a slightly different lens where we'll truly be able to visualize how our design will look under our top-graded surface. That said, in this chapter, we'll review Section View creation, display, and management methods within Civil 3D within the following topics:

- Creating sample lines along alignments
- Creating Section Views to display modeled objects
- Creating intelligent section sheets for plan production

With that, let's go ahead and open up our `Utility Model.dwg` file located within our `Practical Autodesk Civil 3D 2023\Chapter 13\Model` directory. Once it's open, you'll notice that we'll be starting this chapter pretty much where we left off in *Chapter 12, Utility Modeling Tool Belt for Everyday Use*, with the display of our model looking similar to that shown in *Figure 13.1*:

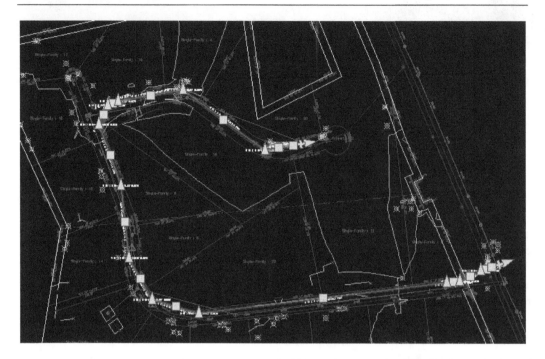

Figure 13.1 – Final appearance of our Residential Subdivision after Chapter 10

Technical requirements

The exercise files for this chapter are available at `https://packt.link/UoiPn`

Creating sample lines along alignments

With our `Utility Model.dwg` file currently open, let's begin familiarizing ourselves with the Cross Section tools available to us within Civil 3D. When we think about Cross Sections, we quite often think about roadway sections as these are where they are most often displayed within a construction drawing plan set. Since we currently have surfaces, corridors, and utilities designed, all of which are currently available within our `Utility Model.dwg` file, we'll want to set ourselves up to include all of these objects in the sampling of the Cross Section sample lines we create; this will become a little clearer in a little bit.

For us to get to the point where we can create Cross Sections, we'll first need to create sample lines along our alignments. These sample lines will be placed so that they are perpendicular to our alignments at specified stations and/or intervals and will be used as a reference for linking our Section Views later on in this chapter.

To create sample lines, we'll go up to our **Home** ribbon and select our **Sample Lines** tool in the **Profile** area and **Section Views** panel. Once activated, we'll be prompted at our command line to select an alignment. Let's go ahead and select our **ALG – Subdivision Main Road – Access** alignment, at which point the **Create Sample Line Group** dialog box will appear, where we'll fill out the fields and make the following selections (also displayed in *Figure 13.2*) and then click the **OK** button:

- **Name**: **SLG – Subdivision Main Road – Access**

- **Description**: **Subdivision access road from York Hwy**

- **Sample Line Style**: **Road Sample Line**

- **Sample Line Label Style**: **Section Station**

- **Select Data Sources to Sample**: Check the **Sample** box next to all the items listed to include all the available ones within our Section Views:

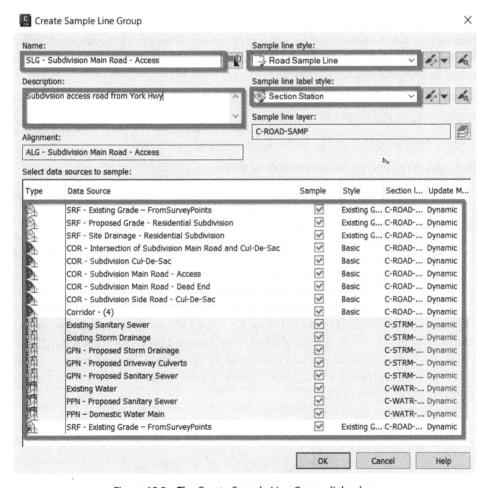

Figure 13.2 – The Create Sample Line Group dialog box

After selecting the **OK** button within the **Create Sample Line Group** dialog box, we'll gain access to our **Sample Line Tools** toolbar. Running from left to right, we have the following tools available to us (also shown in *Figure 13.3*):

1. **Manually Enter a Name for Sample Line**: Allows us to manually enter sample line names as we create them

2. **Edit Name Template**: Allows us to automate the naming convention of sample lines as they're created

3. **Alignment Picker**: Allows us to change alignments by selection as required

4. **Current Sample Line Group**: Lists the current sample line group that we are creating sample lines for

5. **Sample Group Tools**: Allows us to create a new sample line group, edit **Group Defaults**, delete our **Current Group**, select **Group from Drawing**, edit **Swath Widths for Group**, and sample more sources

6. **Sample Line Creation Methods**: Allows us to create sample lines using the following methods: **By Range of Stations, At a Station, From Corridor Stations, Pick Points on Screen**, and **Select Existing Polylines**

7. **Select/Edit Sample Line**: Allows us to select individual entities

8. **Sample Line Entity View**: Changes the view of sample lines to display them in a table format

9. **Undo**: Allows us to undo the previous sample line that was created

10. **Redo**: Allows us to redo the sample line's creation:

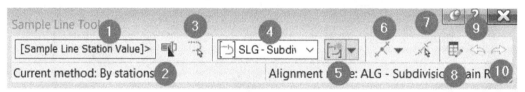

Figure 13.3 – The Sample Line Tools toolbar

With that, let's go ahead and select the **Sample Line Creation Methods** drop-down arrow and select the **By Station Range** option. Once selected, the **Create Sample Lines – By Station Range** dialog box will appear, where we'll make selections as required for the project. I've provided an example of how I would typically approach the sample line creation process if Cross Sections are not required for the project (refer to *Figure 13.4*):

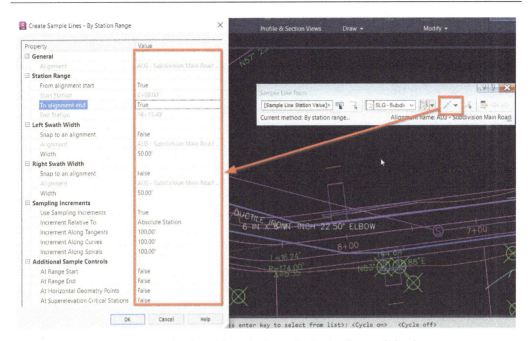

Figure 13.4 – The Create Sample Lines – By Station Range dialog box

Once all the fields have been filled out accordingly, click the **OK** button at the bottom of our **Create Sample Lines – By Station Range** dialog box and watch Civil 3D automatically generate our sample lines per the defined criteria. In our example, we should now be seeing sample lines created along our **ALG – Subdivision Main Road – Access** alignment at **100'** intervals, as shown in *Figure 13.5*:

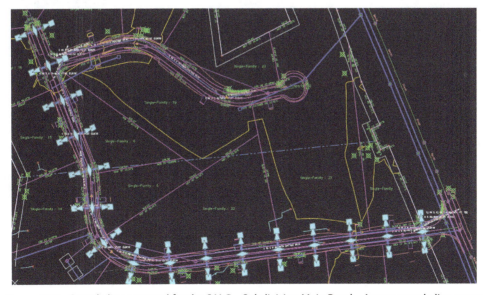

Figure 13.5 – Sample lines created for the SALG – Subdivision Main Road – Access sample line group

With our sample lines and **SLG – Subdivision Main Road – Access** sample line group created for our **ALG – Subdivision Main Road – Access** alignment, let's run through the same workflow for our **ALG – Subdivision Side Road – Cul-De-Sac** alignment.

Jumping back up to our **Home** ribbon, we'll select the **Sample Line** tool in the **Profile** and **Section Views** panel again. In the **Create Sample Line Group** dialog box that appears, we'll fill out the fields make the following selections, and then click the **OK** button:

- **Name: SLG – Subdivision Side Road - Cul-De-Sac**

- **Description: Cul-De-Sac side road connecting to Subdivision Main Road Alignment**

- **Sample Line Style: Road Sample Line**

- **Sample Line Label Style: Section Station**

- **Select Data Sources to Sample**: Check the **Sample** box next to all items listed to include all available within our Section Views

Then, with the **Sample Line Tools** toolbar appearing again, we'll apply the same sample line creation criteria using the **Create Sample Lines – By Station Range** method again, with the final result looking similar to that shown in *Figure 13.6*:

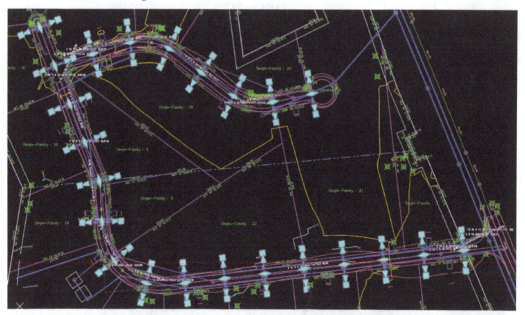

Figure 13.6 – Sample lines created for SLG – Subdivision Side Road – Cul-De-Sac sample line group

Now, with all sample lines, sample line groups, and necessary data sources being sampled, we can go ahead and create data shortcuts for our sample line groups so that we can create our sections in a reference file. We'll want to create these inside a reference file, instead of a model file, so that we can externally reference just the Section Views themselves into our final sheet files when it comes to plan production.

That said, let's pull up our toolspace and activate our **Prospector** tab. Once activated, we'll scroll down to the area where we have our **Data Shortcuts** project listed Right-click the **Data Shortcuts** project name and select **Create Data Shortcuts**. When the **Create Data Shortcuts** dialog box appears, check the boxes next to our newly created sample line groups, **SLG – Subdivision Main Road – Access** and **SLG – Subdivision Side Road – Cul-De-Sac**, and then click **OK** (refer to *Figure 13.7*):

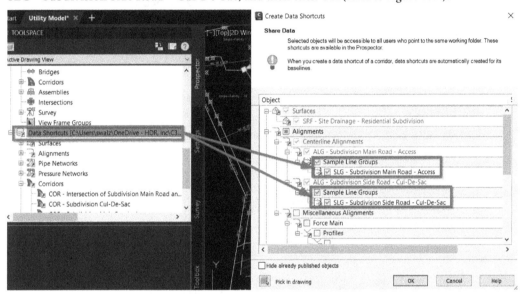

Figure 13.7 – Create Data Shortcuts of Sample Line Groups

Over the next couple of sections, we'll begin to explore how we can carry this information into other files for display and analysis purposes, with the ultimate purpose of preparing our section reference file for eventual incorporation into our construction drawing plan set.

Creating Section Views to display modeled objects

With that, let's go ahead and open up Civil 3D, or go to your start screen if it's already open, and create a new drawing using similar steps outlined in *Chapter 7, Alignments – The Second Foundational Component to Designs within Civil 3D*. We can use our Company Template File.dwt file located in Practical Autodesk Civil 3D 2023\Chapter 13 and select **Open** in the lower right-hand corner of the **Select Template** dialog box. Once our new file has been created, we'll want to save it as Section Reference.dwg in our Practical Autodesk Civil 3D 2023\Chapter 13\Model directory.

As discussed back in *Chapter 3, Sharing Data within Civil 3D*, reference files are intended to contain/ represent 2D geometry and static elements and annotation. Reference files would include content such as surveyed planimetrics, civil site plan geometry, erosion control BMPs, and so on.

Although sections are considered to be Civil 3D objects, only certain elements of sections can be data referenced. In the case of sections, sample lines and sample line groups can be data referenced, but Section Views cannot be, as those are configured in individual files as needed while maintaining a live link to the sample line groups they derive from.

With our `Section Reference.dwg` file created, let's go ahead and jump back into our **Prospector** tab in our toolspace, and then set the **Working Folder** area of our **Data Shortcuts** project to `Practical Autodesk Civil 3D 2023\Chapter 13` location and select the **C3D_2023_123456_Data_ Shortcuts** project, as shown in *Figure 13.8*:

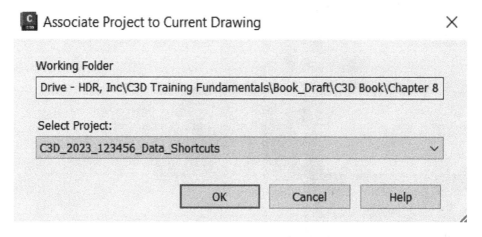

Figure 13.08 – Associate Project to Current Drawing

After our Civil 3D data shortcuts project has been associated with our current file, we can then safely create data references of our **SLG – Subdivision Main Road – Access** and **SLG – Subdivision Side Road – Cul-De-Sac** sample line groups in our `Section Reference.dwg` file.

To do this, we need to expand our **Alignments** category, our **Proposed Conditions** subfolder, then each of our alignments, and finally each of the **Sample Line Groups** categories (refer to *Figure 13.9*), right-click on each of our sample line groups created earlier, and select the **Create…** option:

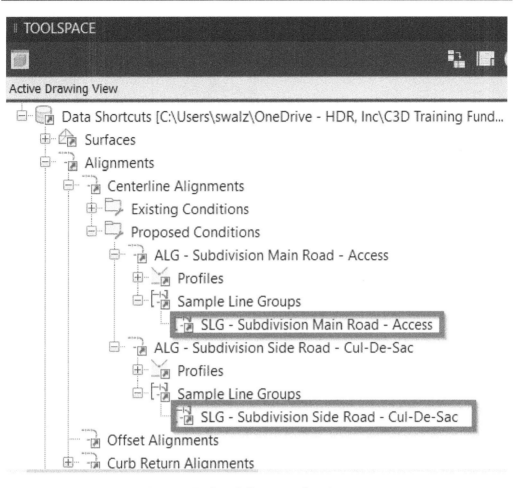

Figure 13.9 – Sample line group data shortcuts

Once created, our **Create Sample Line Group Reference** dialog box will appear. Here, we'll want to jump down to our **Select Data Sources to Sample** section toward the bottom of our dialog box and make sure we have all sources checked, as shown in *Figure 13.10*:

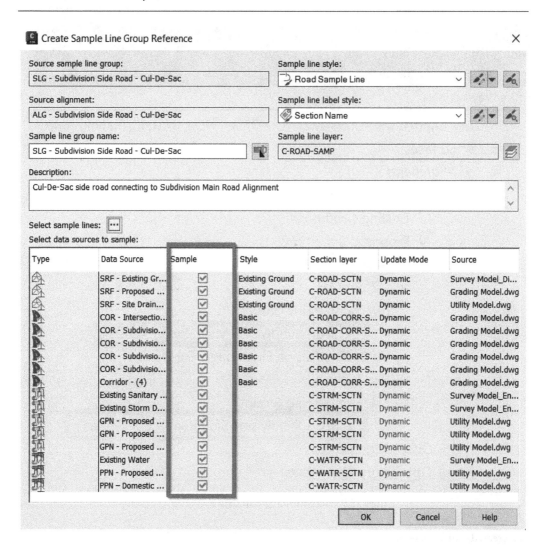

Figure 13.10 – The Create Sample Line Group Reference dialog box

After selecting all data sources, click the **OK** button – Civil 3D will go ahead and import all sample lines and data sources associated with each sample line group. During this importing process, you'll notice that Civil 3D is automatically creating, importing, and linking all data references listed in the **Data Sources** section of our **Create Sample Line Group Reference** dialog box, not just the sample lines and sample line groups.

> **Note**
>
> When we go to select the **Create Reference…** option for the second sample line group, Civil 3D will recognize that those data reference objects are already in the file, so it will not duplicate the insertion of those objects, but rather confirm and insert any that are missing from our current file and link accordingly.

After both sample line groups have been data referenced into our `Section Reference.dwg` file, along with all data sources being sampled, our Model Space should have all reference objects visible, with the output looking similar to what's displayed in *Figure 13.11*:

Figure 13.11 – All data sources referenced in our Section Reference.dwg file

Moving on to creating Section Views, let's go back up to our **Profile and Section Views** panel in our **Home** ribbon and select the down arrow next to **Section Views** (refer to *Figure 13.12*). As shown in *Figure 13.12*, we can quickly see that we have a few different tools available to us related to our Section Views. These tools and their descriptions are as follows:

- **Create Multiple Views**: Allows us to generate and display multiple Section Views within a specified sample line group at one time

- **Create Section View**: Allows us to generate and display a singular Section View within a specified sample line group

- **Project Objects to Multiple Section Views**: Allows us to project objects (AutoCAD Points, Solids, Polylines, and/or Blocks) from the plan view into multiple Section Views

- **Project Objects to Section View**: Allows us to project objects (AutoCAD Points, Solids, Polylines, and/or Blocks) from the plan view into a singular Section View:

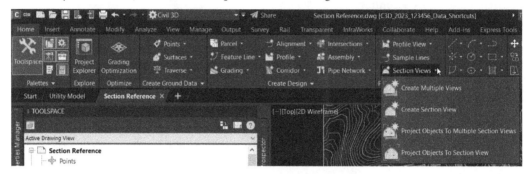

Figure 13.12 – Section Views tools

To give us an idea of the value Section Views can provide, we'll start by using the **Create Section View** option, which will allow us to generate and display a singular Section View within a specified sample line group.

After selecting **Create Section View**, our **Create Section View** dialog box will appear with the **General** tab activated. Using the following steps in the **General** tab, we'll fill out the fields and make selections (also displayed in *Figure 13.13*):

1. **Select Alignment: ALG – Subdivision Main Road – Access**

2. **Sample Line Group Name: SLG – Subdivision Main Road – Access**

3. **Sample Line: 7+00.00**

4. **Section View Name: SCV – Subdivision Main Road – Access**

5. **Description: Section Views along Subdivision Main Road**

6. **Section View Style: Road Section**

7. Click the **Next** button:

Create Section View - General

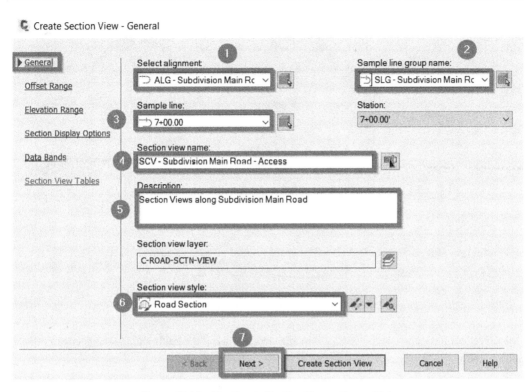

Figure 13.13 – The Create Section View – General tab

In our **Offset Range** tab, we'll use the following steps to fill out the fields and make selections (also displayed in *Figure 13.14*):

1. **Offset Range**: **Automatic** (you'll notice that the **Left** and **Right** values are grayed out. The default values listed are representative of the length of the offset we had associated with our sample lines in the first section.)

2. Click the **Next** button:

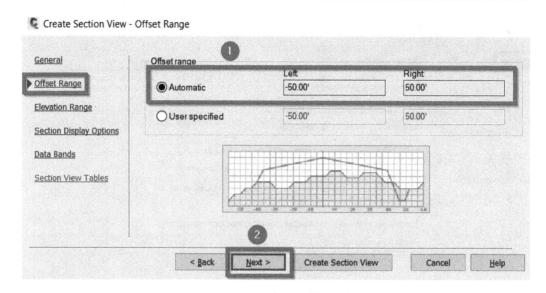

Figure 13.14 – The Create Section View – Offset Range tab

In the **Elevation Range** tab, we'll use the following steps to fill out the fields and make selections (also displayed in *Figure 13.15*):

1. **Offset Range**: **Automatic** (you'll notice that the **Minimum** and **Maximum** values are grayed out. The default values listed are representative of the elevation range that will appropriately display our objects within the Section View we're creating.)

2. Click the **Next** button:

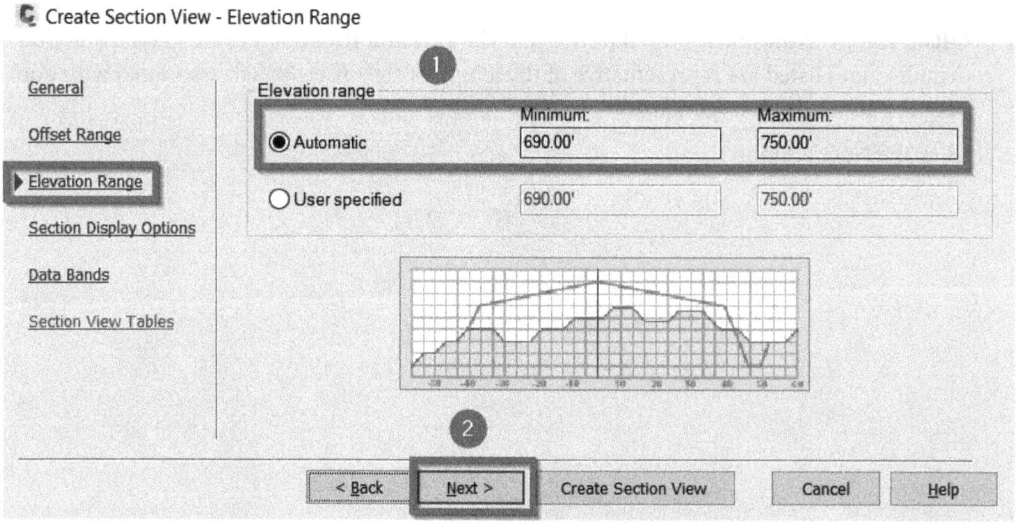

Create Section View - Elevation Range

Figure 13.15 – The Create Section View – Elevation Range tab

In the **Section Display Options** tab, we'll use the following steps to fill out the fields and make selections (also displayed in *Figure 13.16*):

1. Change the **Style** associated with our **SRF – Proposed Grade – Residential Subdivision** Surface Model from **Existing Grade** to **Finished Grade**.

2. Uncheck the **Draw** box associated with our **SRF – Site Drainage – Residential Subdivision** Surface Model (you'll recall that we created this temporary Surface Model in *Chapter 12, Utility Modeling Tool Belt for Everyday Use*, for our Stormwater Analysis).

3. Click the **Next** button:

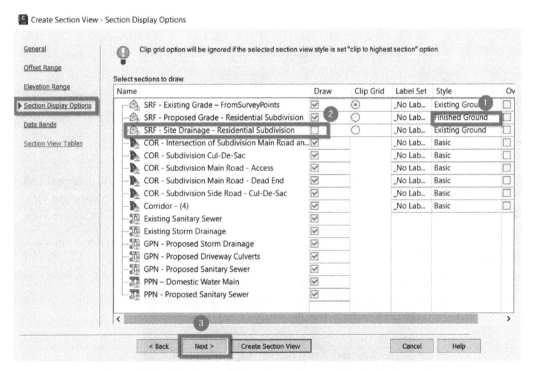

Figure 13.16 – The Create Section View – Section Display Options tab

In the **Data Bands** tab, we'll use the following steps to fill out the fields and make selections (also displayed in *Figure 13.17*):

1. **Select Band Set: Offsets Only**

2. **Location: Bottom of Section View**

3. Click the **Create Section View** button:

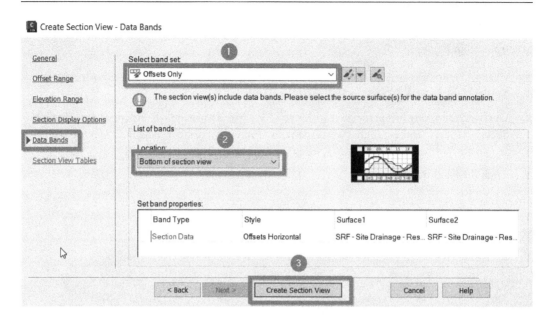

Figure 13.17 – The Create Section View – Data Bands tab

After clicking the **Create Section View** button, the **Create Section View** dialog box will disappear, and we'll be prompted at the command line to identify **Section View Origin**. To identify this **Section View Origin**, we'll click in an open area within our Model Space to manually place our individual Section View.

Now, if we go ahead and select the outer dashed boundary line of our newly created Section View, we'll notice that a few different grips appear on the screen (refer to *Figure 13.18*):

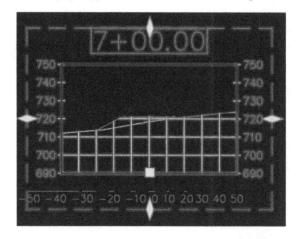

Figure 13.18 – Section View at station 7+00.00

The grips on the top, bottom, left, and right sides of the boundary line allow us to stretch our **Drafting Buffer Outline**. This **Drafting Buffer Outline** does not provide much value when generating singular Section Views but can have an effect when we create multiple Section Views for plan production purposes.

The boxed grip at the bottom of our Section Grid represents the actual insertion point of our Section View. If we click on this grip, we can move/relocate the Section View anywhere else that we'd like, still while maintaining that dynamic link to the data sources that are being displayed.

Now that we've been able to get an idea of what the singular Section View generation workflow looks like, let's move on to creating multiple Section Views that we will be able to incorporate into our plan production set and further analyze all of our modeled objects within our Residential Subdivision design.

Creating intelligent section sheets for plan production

Creating multiple Section Views has a few different benefits and advantages, with the biggest one being that we can easily configure our Section Views so that they are already prepared for construction drawing sheeting. A few more additional benefits are as follows:

- Dynamic linking to sample lines and sample line groups
- Adjustable sheet drafting buffer outlines that will dynamically update all remaining Section Views
- Quickly analyze all modeled objects in all Section Views

With that, let's go up to our **Profile and Section Views** panel in our **Home** ribbon and select the **Create Multiple Views** tool. Once selected, the **Create Multiple Section Views** dialog box will appear. Right off the bat, you'll notice that we have quite a few more options and selections to make than what was available in the previous exercise when we were creating just a singular Section View.

Starting with our **General** tab activated within the **Create Multiple Section Views** dialog box, we'll fill out the fields and make selections as follows (also displayed in *Figure 13.19*) and click the **Next** button:

- **Select Alignment**: **ALG – Subdivision Main Road – Access**
- **Sample Line Group Name**: **SLG – Subdivision Main Road – Access**
- **Station Range**: **Automatic**
- **Section View Name**: **SCV – Subdivision Main Road – Access**
- **Description**: **Section Views along Subdivision Main Road**
- **Section View Style**: **Road Section**:

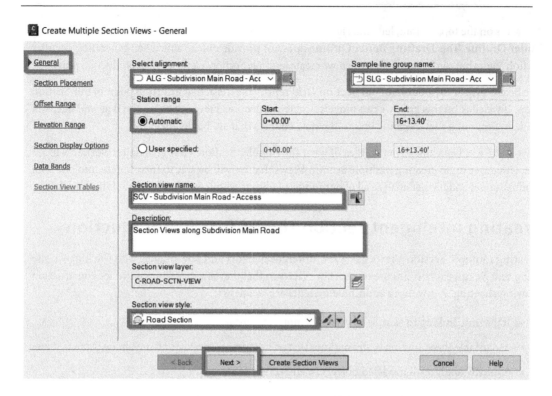

Figure 13.19 – The Create Multiple Section View – General tab

Next, in the **Section Placement** tab, we'll use the following steps to fill out the fields and make selections (also displayed in *Figure 13.20*) and then click the **Next** button:

1. Select the **Production – Use a layout from a template file .dwt to place sections on sheets** option.

2. Click on the ellipsis button to specify a template and layout to create new section sheets.

3. In the **Select Layout as Sheet Template** dialog box, click on the ellipsis button to specify a template and layout to create new section sheets.

4. Select the corresponding sheet layout that we wish to use for plan production purposes. In our case, we'll opt for **ANSI D Section 40 Scale**.

5.   Click on the **OK** button in the **Select Layout as Sheet Template** dialog box.

6.   Select **Basic** for **Group Plot Style**:

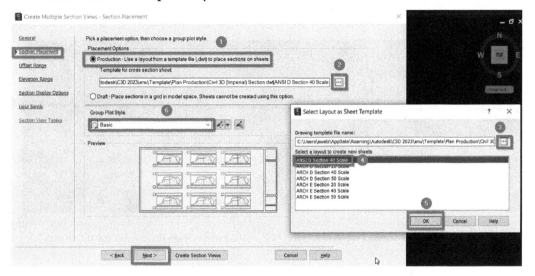

Figure 13.20 – The Create Multiple Section View – Section Placement tab

> **Note**
>
> Note that we were to select the **Draft – Place sections** option in a grid in Model Space, sheets cannot be created using this option, so Civil 3D will place all the Section Views in a grid. These Section Views are great for quickly analyzing all Civil 3D modeled objects that cross our sample lines but can be difficult to organize for sheeting purposes later on down the road. This is not to say it can't be done, but it makes it more time-consuming if there is a need to sheet all Section Views.

The final four tabs (**Offset Range**, **Elevation Range**, **Section Display Options**, and **Data Bands**) are the same as what we previously reviewed when we were generating individual Section Views. That said, we'll apply the same selections we had in the previous section and click on the **Create Section Views** button at the bottom of the **Create Multiple Section Views** dialog box.

After selecting the **Create Section View** button, the **Create Section View** dialog box will disappear, and we'll be prompted at the command line to **Identify Section View Origin**. Just as we did earlier, we'll click in an open area within our Model Space to manually place our Section Views, with our multiple section view layout looking similar to that shown in *Figure 13.21*.

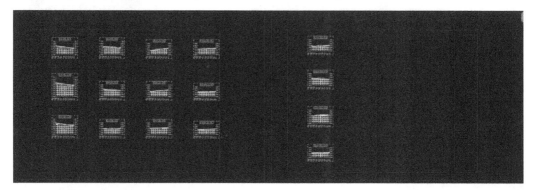

Figure 13.21 – Multiple section views along the ALG – Subdivision Main Road – Access alignment

With all Section Views created along our **ALG – Subdivision Main Road – Access** alignment, let's run through the same process for our sample line group associated with **ALG - Subdivision Side Road - Cul-De-Sac** alignment. We'll want to apply the same steps as before, but with the following changes specified in our **General** tab within the **Create Multiple Section Views** dialog box:

1. **Select Alignment**: **ALG – Subdivision Side Road – Cul-De-Sac**
2. **Sample Line Group Name**: **SLG – Subdivision Side Road – Cul-De-Sac**
3. **Section View Name**: **SCV – Subdivision Side Road – Cul-De-Sac**
4. **Description**: **Section Views along Subdivision Side Road – Cul-De-Sac**

After running through all the steps within the **Create Multiple Section Views** dialog box and then placing our multiple Section Views in an open area within our Model Space, we can start looking at different ways to analyze the sections that have been generated for us.

The first and most obvious way to analyze our sections is by simply looking at/reviewing each Section View to verify that all Civil 3D modeled objects displayed make sense (that is, gravity and pressure network piping are located below the finished grade, proposed grade lines are tying back into existing grade lines as expected, corridor model geometry appears as expected, and so on).

Quick tip

If we select any one of our Section Views to pull up the Section View **Contextual** ribbon, we'll gain access to the **Analyze** panel, which contains a **Station Tracker** tool. If we have our Model Space view set up with multiple views, where one view is displaying Section Views while another is displaying our Plan View, we can quickly locate where each section is located within our Plan View, as shown in *Figure 13.22*:

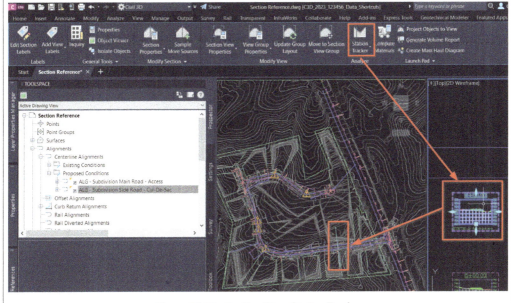

Figure 13.22 – Section View Station Tracker

The second type of analysis that can be performed at this stage allows us to generate quantity reports of material and analyze cut versus fill as it relates to earthwork so that we can better gauge how much dirt needs to be moved/relocated and either carried off-site during construction or brought in to build our site up.

All these types of metrics will provide significant value and insight to contractors as they are bidding for the job and provide enough information for the owner to make a more informed decision as it relates to funding. These tools are available through the **Compute Materials** tool available in the **Launch Pad** panel within our Section View **Contextual** ribbon, as shown in *Figure 13.23*:

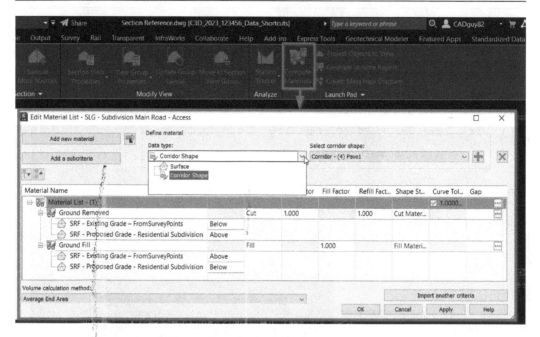

Figure 13.23 – Section View Station Tracker

Once we're satisfied with how our sections look and that all modeled objects throughout our Residential Subdivision design are accurately located and in conformance to design standards and regulations, we can confidently say that we are finished with our design.

More and more frequently, we are seeing clients and agencies requesting a digital delivery solution, whereas designs are submitted with **Models as the Legal Document (MALD)**, and plan production sheets are for information only. That said, at present, there is still a need to take the next steps to finalize all reference files in preparation for our construction drawing plan set.

Summary

By working through this chapter, we officially completed our Residential Subdivision design. We also gained the assurance that all modeled objects are located where they should be and as we intended them to be in conformance with design standards and regulations. We learned how we can generate Section Views (should the project require them) and maintain a dynamic link to all modeled objects throughout our design where our sample lines intersect. Furthermore, we also learned how we can leverage Section Views to further analyze our modeled objects and overall design.

In the next chapter, we'll take a look at the plan production capabilities and tools available to us within Civil 3D, where we can automatically generate sheet files of our design. This will essentially be the final step that we'll want to take just before submitting our design as part of the digital delivery process. Historically, this effort is begun much earlier on in the design process, typically just before a conceptual design review.

However, as the modeling process becomes more intricate, and technology advancements continue to increase the possibilities of reviewing our models in place of sheets, we must spend more time and effort up front, ensuring that our models are truly representative of our design and that we are maintaining dynamic linking between as many modeled objects as possible to minimize risk and rework later on down the road.

14

Automating Sheet Creation

Up to this point, we've taken all steps necessary to generate a Residential Subdivision design that we feel comfortable with, all while learning all of the design and analysis tools that are available to us within Civil 3D. We've also gained the assurance that all modeled objects are located where they should be and as we intended to conform to design standards and regulations.

With our Residential Subdivision design all modeled and wrapped up, we can now move on to understanding how we can leverage some automated workflows to set up our construction drawings by using Civil 3D's Plan Production tools. Utilizing these tools will provide a new level of consistency and efficiency to us throughout the entire project design life cycle with its dynamic capabilities. This will essentially be the final step that we'll want to take before submitting our design as part of the digital delivery process.

Historically, this effort has begun much earlier on in the design process, typically just before a Conceptual Design review. However, as the modeling process becomes more intricate, and technology advancements continue to increase the possibilities of reviewing our models in place of sheets, we must spend more time and effort upfront ensuring that our models are truly representative of our design and that we maintain dynamic linking between as many modeled objects as possible to minimize risk and reworking later on down the road.

All that said, in the chapter, we'll cover the following areas:

- Automating Plan Sheet creation
- Automating Plan and Profile Sheet creation
- Automating Cross Section Sheet creation

With that, let's go ahead and open up Civil 3D, or go to the **Start** screen if it's already open, and create a new drawing using similar steps to those outlined in *Chapter 7, Alignments –The Second Foundational Component to Designs within Civil 3D*. We can use our Company Template File.dwt file located in Practical Autodesk Civil 3D 2023\Chapter 14 and select **Open** in the lower right-hand corner of the **Select Template** dialog box. Once our new file has been created, we'll want to save it as a Site Plan Sheets.dwg file to Practical Autodesk Civil 3D 2023\Chapter 14\Sheet.

As discussed back in *Chapter 3, Sharing Data within Civil 3D*, Sheet files are the final product of the project, where the construction of the Sheet files will consist of externally referencing both the Reference and Model files, along with sheet borders, general notes, north arrows, and any additional sheet specific annotation.

With that, let's go ahead and attach our Survey Model.dwg as an Overlay, contained within our Practical Autodesk Civil 3D 2023\Chapter 14\Model location, as well as Site Plan Reference.dwg as an Overlay, contained within Practical Autodesk Civil 3D 2023\Chapter 14\Reference.

Then, we'll want to jump back into the **Prospector** tab in our Toolspace, and then set the **Working Folder** option in our Data Shortcuts project to Practical Autodesk Civil 3D 2023\Chapter 14 location and select the C3D_2023_123456_Data_Shortcuts project, as shown in *Figure 14.1*.

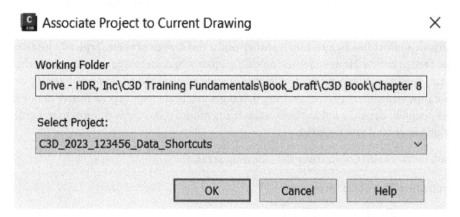

Figure 14.1 – Associating the current drawing with the project

After our Civil 3D Data Shortcuts project has been associated with our current file, we can then safely create data references as needed later on down the road.

Then, with our Site Plan Sheet.dwg file set up for us to include all objects necessary to represent and reference the existing and proposed built environment, we should have a plan view that looks similar to that shown in *Figure 14.2*.

Figure 14.2 – Site Plan Sheets.dwg file

With our `Site Plan Sheets.dwg` file configured, let's go ahead and explore ways we can leverage Civil 3D's tools to automate the way we create or generate our sheets and overall construction drawing set.

Technical requirements

The exercise files for this chapter are available at `https://packt.link/UoiPn`

Automating Plan Sheet creation

With our `Site Plan Sheets.dwg` file currently open, let's begin familiarizing ourselves with the **Map Task Pane** tools available to us within Civil 3D. The **Map Task Pane** tools create, manage, display and publish maps, and can be accessed from the **Palettes** pulldown in our **Home** ribbon and by selecting the **Map Task Pane** tools as shown in *Figure 14.3*.

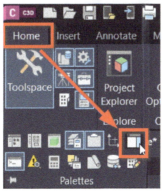

Figure 14.3 – Activating Civil 3D's Map Task Pane tools

Once activated, you'll receive a prompt at the Command Line asking you whether you'd like the task pane **On** or **Off**, at which point we'll select (or type) **On** and hit the *Enter* key on our keyboard.

Using the following steps, we'll be able to automate the creation of our Plan sheets to be included in our construction drawing set (also displayed in *Figure 14.4*):

1. When the **Map Task Pane** appears, let's go ahead and select the **Map Book** tab.

2. Next, we'll select **New** in the top-left corner of our **Map Book** tab.

3. Finally, we'll select **Map Book** to create a new Map Book.

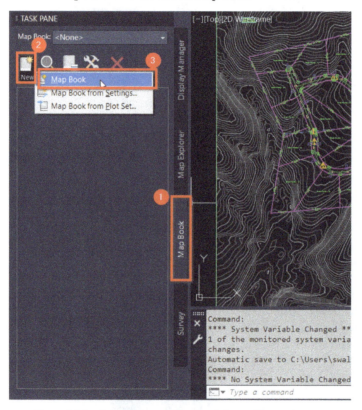

Figure 14.4 – Creating a new Map Book in our Map task pane

After running through those steps, the **Create Map Book** dialog box will then appear, where we'll make the following selections (also displayed in *Figure 14.5*):

*   **Source: Model Space**

*   **Sheet Template: Settings**

*   **Tiling Scheme: By Area**

*   **Naming Scheme: Grid Sequential**

- **Key**: **Layers**

- **Legend**: **None**

- **Sheet Set**: **Create New**

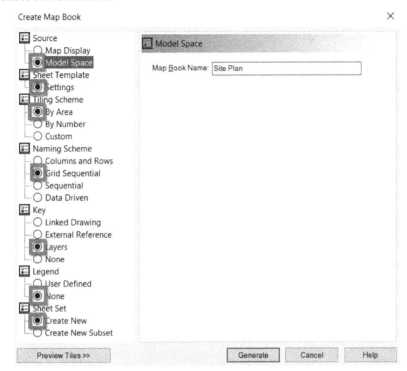

Figure 14.5 – Initial selections in Map Book

4. Next, we'll want to go a bit deeper into each of our selections to fill out available fields and options as follows:

- **Source**: **Model Space**

- **Map Book Name**: `Site Plan`

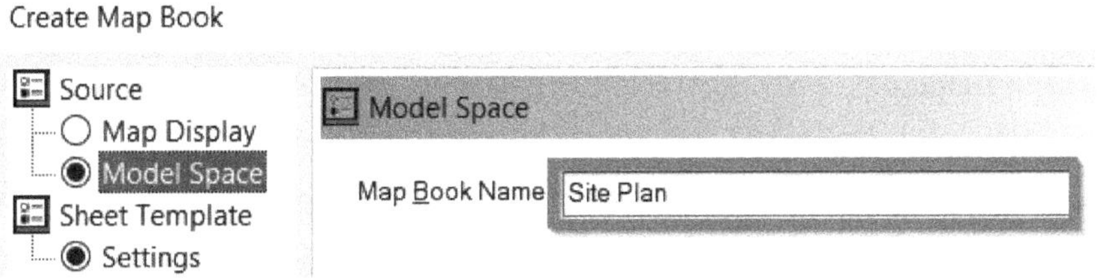

Create Map Book

Figure 14.6 – Map Book > Source > Model Space Criteria

- **Sheet Template: Settings** (as shown in *Figure 14.7*):

    - **Choose a Sheet Template**: `C:\Users\username\AppData\Roaming\Autodesk\ C3D 2023\enu\Template\AutoCAD Template\Map Book Template - 22x34 Classic.dwt`

    - **Choose a Layout: Ansi D**

    - **Include a Title block (name or file): Title Block**

    - **Include Adjacent sheet links (name or file): Adjacent_Arrow_4**

    - **Scale Factor:** `20`

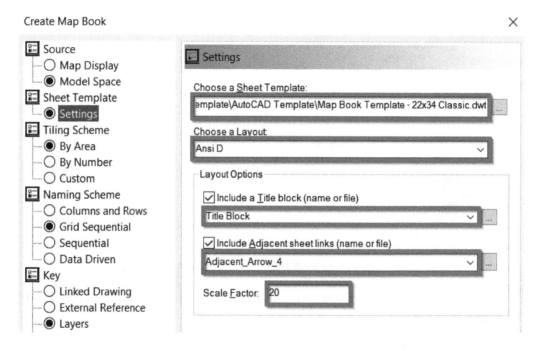

Figure 14.7 – Map Book > Sheet Template > Settings Criteria

- **Tiling Scheme: By Area** (as shown in *Figure 14.8*):

    - **Layer: Defpoints**

    - **First corner:** `1859517.48` and `1188684.05`

    - **Opposite corner:** `1861121.99` and `1190000.02`

    - **% overlap of each tile:** `0`

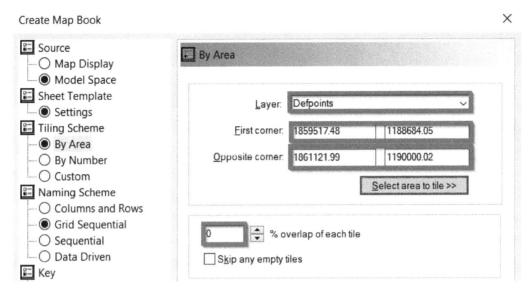

Figure 14.8 – Map Book > Tiling Scheme > By Area Criteria

> **Note**
> In lieu of plugging in coordinates for **First corner** and **Opposite corner**, we can alternatively use the **Select area to tile >>** button. When this has been activated, we'll be able to draw a window around the extent of our project site and/or the area that we'd like to include in the automatic Plan Sheet creation process.

- **Naming Scheme: Grid Sequential** (as shown in *Figure 14.9*):

  - **Begin with: Rows**

  - **Order from: Upper Left to Lower Right**

  - **Start with:** 1

  - **Increment by:** 1

  - **Keep names for skipped tiles:** *Checked*

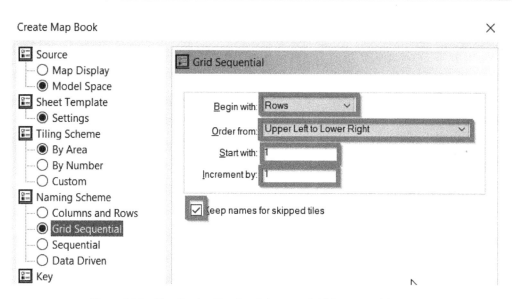

Figure 14.9 – Map Book > Naming Scheme > Grid Sequential Criteria

- **Key: Layers** (as shown in *Figure 14.10*). Select **All Layers** (easiest to use by pressing *Shift*) and click on the **Add Layers for Map Key** button.

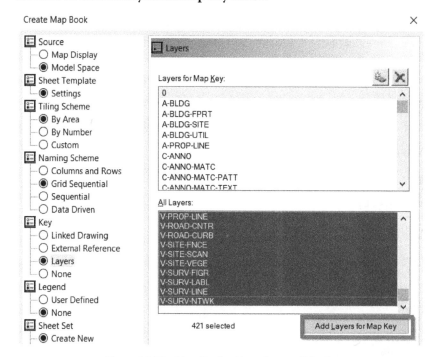

Figure 14.10 – Map Book > Key > Layers Criteria

- **Legend**: **None** (as shown in *Figure 14.11*):

  - **Sheet Set**: **Create New**:

    - In the path, select `Practical Autodesk Civil 3D 2023\Chapter 14\ Residential Subdivision.dst`

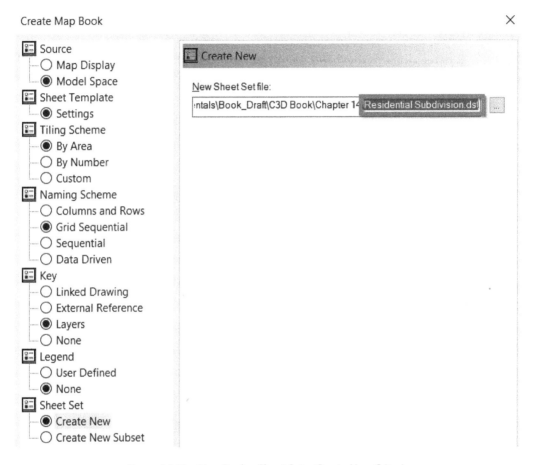

Figure 14.11 – Map Book > Sheet Set > Create New Criteria

5. Finally, with all of our Map Book settings in place, let's go ahead and click the **Generate** button at the bottom of the **Create Map Book** dialog box and watch the magic happen. Shortly after clicking on the **Generate** button, we'll first notice that we now have a gridded layout displayed in Model Space covering our entire Residential Subdivision layout, as shown in *Figure 14.12*.

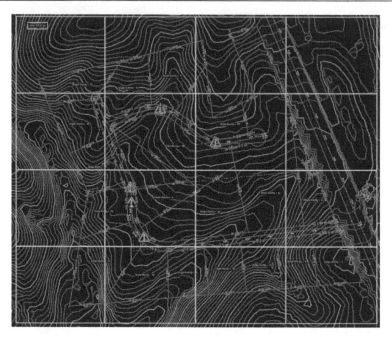

Figure 14.12 – Site Plan Sheets.dwg Model Space Gridded Layout

The second, and more important, item that we'll notice is that along the bottom of our Civil 3D session, we now have 16 new Layouts created from the newly gridded layout of our design, as shown in *Figure 14.13*.

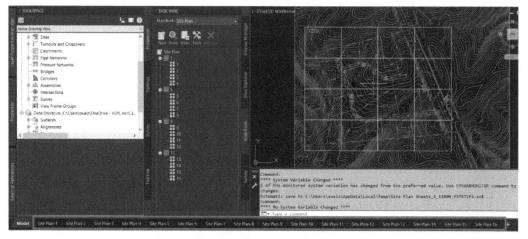

Figure 14.13 – Site Plan Sheets.dwg Layout tabs

As we flip through and activate each of our **Layout** tabs, we'll see how easy Civil 3D's Map Books can be at automating our first run at Plan Sheet generation. If, for whatever reason, we need to revisit our layout, we can always go back to our Map Book in **Map Task Pane**, right-click with our mouse on **Site Plan Map Book**, and select the **Edit Settings and Rebuild…** option, as shown in *Figure 14.14*.

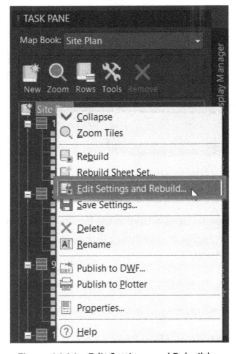

Figure 14.14 – Edit Settings and Rebuild…

Now that we're satisfied with the **Site Plan** layouts to be included in our overall construction drawing set, we can use similar processes and tactics to generate additional plan-only sheets that have the potential to display Grading Plans, Erosion Control Plans, Utility Plans, and so on.

Automating Plan and Profile Sheet creation

Once we've gone ahead and created all plan-only Layouts and Sheets to be included in our construction drawing set, we can take the next step in automating the generation of our Plan and Profile Sheets. To automate the creation of these types of sheets, we'll want to take a slightly different approach.

To get started, let's go ahead and open up our Utility Model.dwg file located in Practical Autodesk Civil 3D 2023\Chapter 14\Model. The reason we're starting with this Utility Model.dwg file instead of creating a new file to begin our Plan and Profile sheet creation is due to the fact that Civil 3D's Plan Production Tools require a file that contains all anticipated modeled geometry that we'd like to include in our Plan and Profile sheets to be present in the current drawing.

Additionally, the automatic sheet creation process in using the Plan Production tools will automatically create a new drawing file for each Plan and Profile sheet we are attempting to generate.

Once we have opened the Utility Model.dwg file, let's navigate up to our ribbons along the top, select the **Output** ribbon, and then select the **Create View Frames** tool within the **Plan Production** Panel, as shown in *Figure 14.15*.

Figure 14.15 – Activating the Create View Frames tool

Once the **Create View Frames** tool has been activated, the **Create View Frames** dialog box will appear. You'll notice that there are several tabs, or selections, that require input along the left-hand side of the **Create View Frames** dialog box.

Running through them, from top to bottom, we'll fill out fields and makes selections as follows:

1. **Alignment** (as displayed in *Figure 14.16*):

    - **Alignment**: ALG - Subdivision Main Road - Access

    - **Station Range**: Automatic

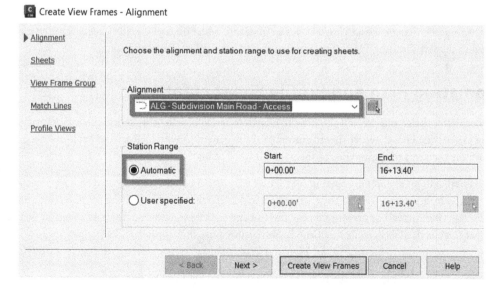

Figure 14.16 – Create View Frames dialog box – the Alignment tab

2. **Sheets** (as displayed in *Figure 14.17*):

    - **Sheet Settings: Plan and Profile**

    - **Template for Plan and Profile Sheet**: `C:\Users\username\AppData\Roaming\Autodesk\C3D 2023\enu\Template\Plan Production\Civil 3D (Imperial) Plan and Profile.dwt|ANSI D Plan and Profile 40 Scale`

    - **View Frame Placement: Along Alignment**

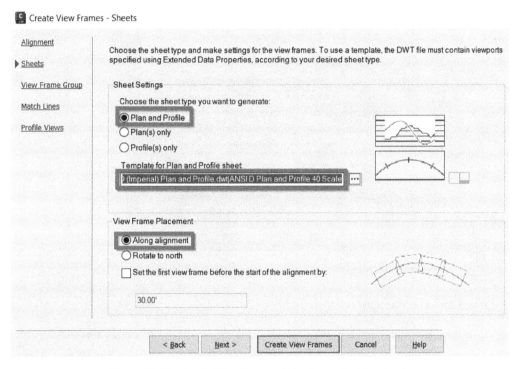

Figure 14.17 – Create View Frames dialog box – the Sheets tab

3. **View Frame Group** (as displayed in *Figure 14.18*):

    - **Name: Plan and Profile - Subdivision Main Road - Access**

    - **Description: Plan and Profile View Frames displayed along Subdivision Main Road - Access Alignment**

    - **Layer**: `C-ANNO-VFRM`

    - **Name**: `VF - (<[Next Counter(CP)]>)`

    - **Style: Basic**

- **Label style: Basic**

- **Label location: Top left**

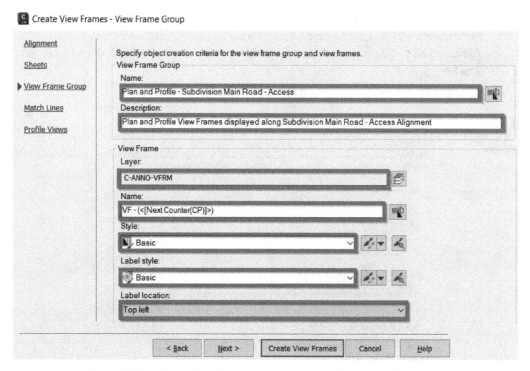

Figure 14.18 – Create View Frames dialog box – the View Frame Group tab

4. **Match Lines** (as displayed in *Figure 14.19*):

- **Snap station value down to the nearest:** *Check the box* and set **Station Value** to 50

- **Layer:** C-ANNO-MTCH

- **Name:** ML - (<[Next Counter(CP)]>)

- **Style: Basic**

- **Left label style: Basic Left**

- **Right label style: Basic Right**

- **Left label location: End**

- **Right label location: Start**

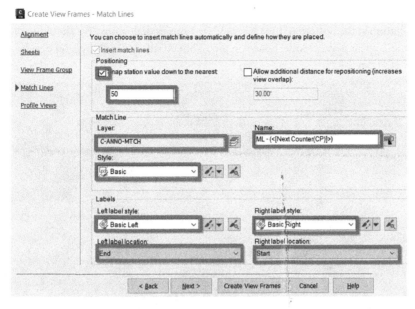

Figure 14.19 – Create View Frames dialog box – the Match Lines tab

5. **Profile Views** (as displayed in *Figure 14.20*):

- **Select profile view style: Profile View**

- **Select band set style: Plan Profile Sheets - Elevations and Stations**

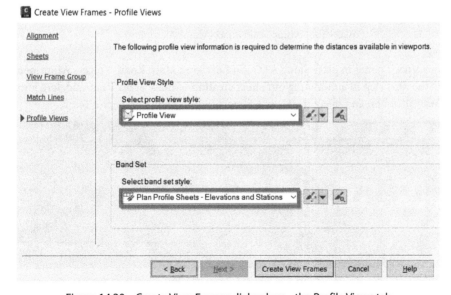

Figure 14.20 – Create View Frames dialog box – the Profile Views tab

After filling out all fields and making the selections as detailed, let's go ahead and click on the **Create View Frames** button at the bottom of the **Create View Frames** dialog box. We should now be seeing two View Frames appear in our Model Space on top of our Residential Subdivision design, similar to that displayed in *Figure 14.21*.

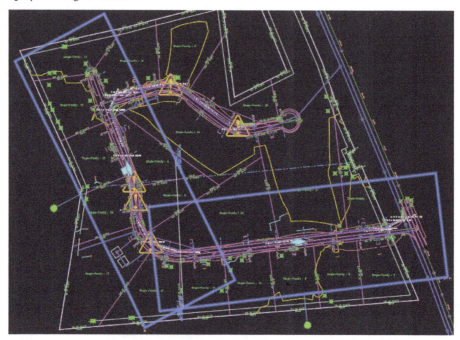

Figure 14.21 – View Frames added to Model Space along our
ALG - Subdivision Main Road -Access Alignment

Now, with our View Frames in place along **ALG - Subdivision Main Road - Access Alignment**, we'll want to take the next step in automating our sheet creation of these areas. That said, let's jump back up to our **Output** ribbon and select the **Create Sheets** tool, as shown in *Figure 14.22*.

Figure 14.22 – Activating the Create Sheets tool

Once the **Create Sheets** tool has been activated, the **Create Sheets** dialog box will appear. Similar to the **Create View Frames** dialog box, you'll notice that there are several tabs, or selections, that require input along the left-hand side of the **Create Sheets** dialog box. Running through them, from top to bottom, we'll fill out fields and make selections as follows:

1. **View Frame Group and Layouts** (as displayed in *Figure 14.23*):

    -   **View Frame Group: Plan and Profile - Subdivision Main Road - Access**

    -   **View frame range: All**

    -   **Layout Creation: All layouts in one new drawing**

    -   **Layout Name: Plan and Profile - Subdivision Main Road - Access**

    -   **Choose the north arrow block to align in layouts: North**

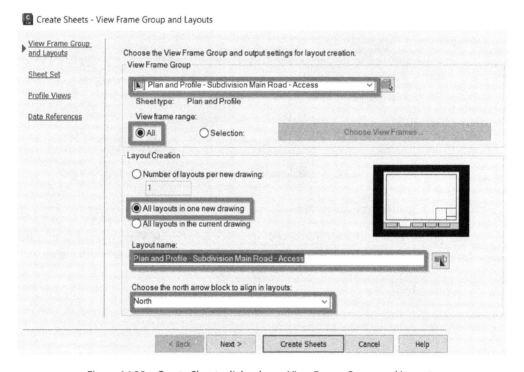

Figure 14.23 – Create Sheets dialog box – View Frame Group and Layouts

2. **Sheet Set** (as displayed in *Figure 14.24*):

    -   **Sheet Set**: **Add to existing sheet set** and name it as `Residential Subdivision`

    -   **Sheet files storage location**: `C:\Users\swalz\OneDrive - HDR, Inc\C3D Training Fundamentals\Book_Draft\C3D Book\Chapter 14\Sheet\`

- **Sheet file name**: **Plan and Profile Sheets**

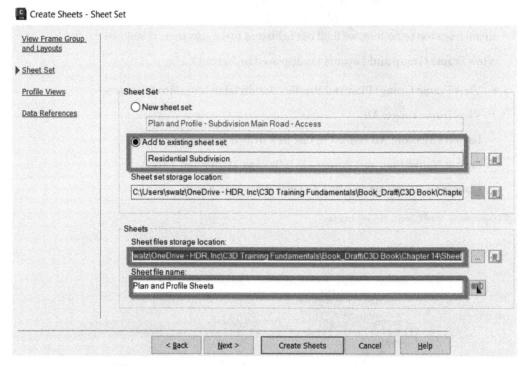

Figure 14.24 – Create Sheets dialog box – Sheet Set

3. **Profile Views** (as displayed in *Figure 14.25*):

    - **Other profile view options**: **Choose settings** and select **Profile View Wizard …** to make any adjustments necessary in the **Create Multiple Profile Views** dialog box that appears

    - **Align views**: **Align profile and plan view at start**

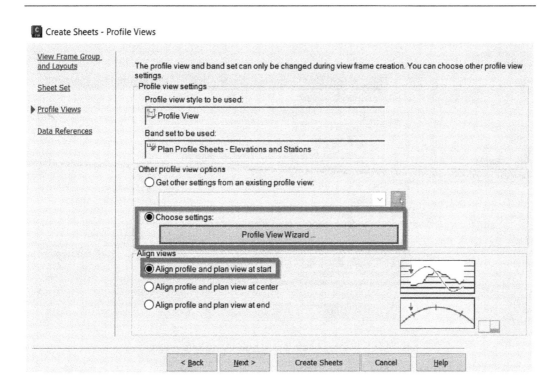

Figure 14.25 – Create Sheets dialog box – Profile Views

4. **Data References** (as displayed in *Figure 14.26*):

    -   **Select the data you want referenced in your sheets**: Select either individual or all grouped modeled objects as appropriate

    -   **Copy pipe network labels to destination drawings**: *Check the box*

    -   **Copy pressure network labels to destination drawings**: *Check the box*

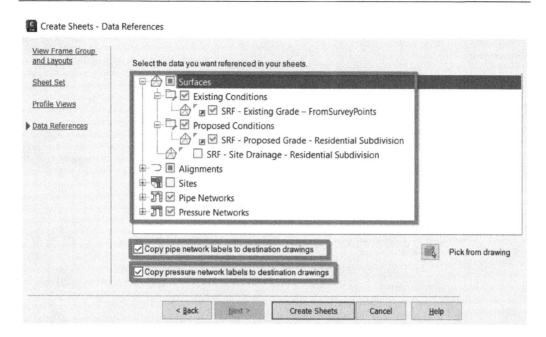

Figure 14.26 – Create Sheets dialog box – Date References

5. With all of our **Create Sheets** settings defined, let's go ahead and click on the **Create Sheets** button at the bottom of the **Create Sheets** dialog box. After clicking **Create Sheets**, we'll be prompted to select **Profile View Origin** at the Command Line.

6. Go ahead and click in a blank area in our Model Space to place the new Profile Views. After defining the location, we can now sit back and watch the magic happen where Civil 3D will automatically create our sheets and add them to our **SHEET SET MANAGER** list (as shown in *Figure 14.27*).

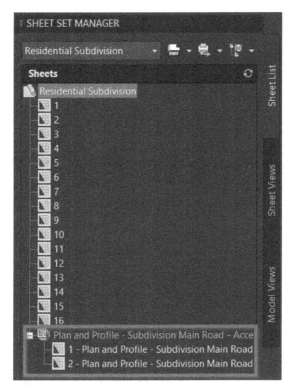

Figure 14.27 – Updated Sheet Set Manager

With our first run-through of automating the creation of our Plan and Profile sheets under our belt, let's go ahead and run through similar processes to be applied to our **ALG - Subdivision Side Road - Cul-De-Sac** Alignment. A lot of the same fields and selections will be similar in both our **Create View Frame** and **Create Sheets** dialog boxes. We'll quickly realize that due to the scale (1" = 40') being applied and the short length of our **ALG – Subdivision Side Road – Cul-De-Sac** Alignment, only one sheet will actually be created.

In any event, this practice and workflow are still advisable to drive consistency, leverage automation as much as possible, and keep all newly created Sheets available in our Sheet Set Manager. After running through the process Plan and Profile Sheet Creation process again, let's jump over to our Section Sheets to gain some insight as to how we can continue to leverage automated sheet generation workflows.

Automating Cross Section Sheet Creation

Now that we have created our plan-only and Plan and Profile Sheets using the various automated methods available to us within Civil 3D, let's go ahead and place our Cross Sections into Sheets using automated workflows as well. To automate the creation of these types of sheets, we'll (yet again) want to take a slightly different approach.

That said, let's start by opening up our `Section Reference.dwg` file located in `Practical Autodesk Civil 3D 2023\Chapter 14\Reference`. After opening, we'll want to immediately use **Save As** on `Section Sheets.dwg`, and place it in our `Practical Autodesk Civil 3D 2023\Chapter 14\Sheet` location. Using **Save As** allows us to take what we've already developed in this file and save it as a Sheet File for our automated Section sheet generation.

> **Note**
>
> Later on, we'll end up deleting all contents from Model Space in our newly created Sheet File and replacing it by externally referencing in our `Section Reference.dwg`. Speaking from experience, although this workflow sounds a little backward, it's actually the most foolproof process we have available to us at present time with Civil 3D.

With our new `Section Sheets.dwg` file created and currently open, let's navigate back up to our **Output** ribbon along the top of our Civil 3D session, and select the **Create Section Sheets** Tool within our **Plan Production** Panel, as shown in *Figure 14.28*.

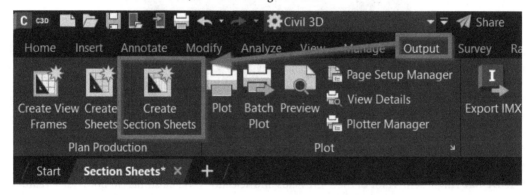

Figure 14.28 – Activating the Create Section Sheets tool

When the **Create Section Sheets** tool has been activated, the **Create Section Sheets** dialog box will appear, where we'll use the following steps to fill out the fields and make necessary selections and generate our Cross Section Sheets (also displayed in *Figure 14.29*):

1. **Select alignment**: **ALG - Subdivision Main Road - Access**

2. **Sample line group name**: **SLG - Subdivision Main Road - Access**

3. **Select section view group**: **Section View Group - 1**

4. **Layout name**: `Cross Sections - (<[Next Counter(CP)]>)`

5. **Sheet Set**: Select the **Add to existing sheet set** option, click on the ellipsis, navigate to `Practical Autodesk Civil 3D 2023\Chapter 14\`, and select **Residential Subdivision**.

6. Select the **Create Sheets** button.

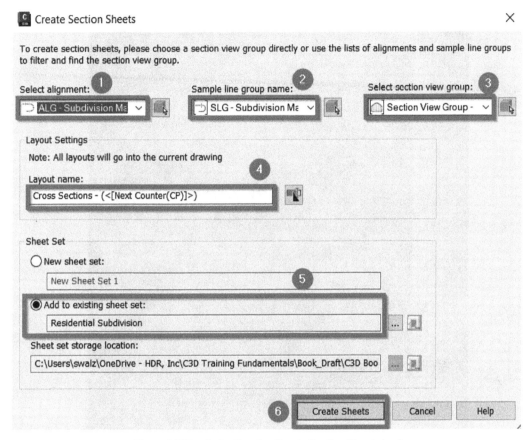

Figure 14.29 – Activating the Create Section Sheets tool

After selecting the **Create Sheets** button in our **Create Section Sheets** dialog box, we can sit back (again) and watch (more of) the magic happen where Civil 3D will automatically create our sheets and add them to our **SHEET SET MANAGER** list (as shown in *Figure 14.30*).

Figure 14.30 – Updated Sheet Set Manager

We'll also notice that we now have two new Layouts created along the bottom of our drawing (refer to *Figure 14.31*), which are our official Cross Section sheets that will be included in our overall construction drawing set for submission.

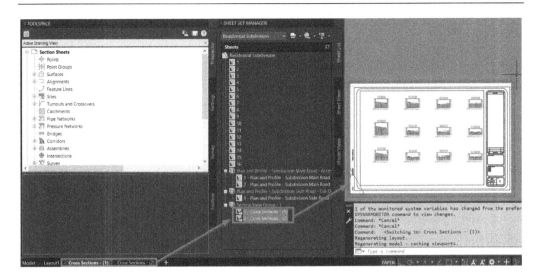

Figure 14.31 – New Layouts added to Cross Section Sheets.dwg

With our first run-through of automating the creation of our Cross Section Sheets under our belt, let's go ahead and run through similar processes to be applied to our **ALG - Subdivision Side Road - Cul-De-Sac** Alignment. A lot of the same fields and selections will be similar in both our **Create Section Sheets** dialog boxes.

As was the case with our Plan and Profile Sheet creation process, we'll notice that due to the shorter length of our **ALG - Subdivision Side Road - Cul-De-Sac** Alignment, only one sheet will actually be created.

After all the Cross Section Sheets have been created, we can then safely go back into Model Space and replace the existing content with that of our `Section Reference.dwg` file located in `Practical Autodesk Civil 3D 2023\Chapter 14\Reference`.

To do so, we'll take the following steps:

1. Select **All objects** (*Ctrl + A*).

2. Delete/erase all selected objects.

3. Enter the `XREF` command.

4. Attach the `Section Reference.dwg` file located in `Practical Autodesk Civil 3D 2023\Chapter 14\Reference` as an Overlay.

Now, with all Sheet Files showing Plan only, Plan and Profile, and Sections, we've been able to shore up most of our construction drawing set. Obviously, this doesn't necessarily complete our construction drawing set as we still need to pull together our Cover, General, and Detail sheets, but following these workflows will essentially get you close to 80% of the way there.

Summary

As we worked through this chapter, we experienced firsthand how the proper configuration and linking the modeled objects included in our design will allow us to leverage automation in the end.

With Civil 3D's ability to dynamically link and associate a multitude of model objects together, we are essentially eliminating necessary rework and the risk of going over budget and schedule, as we have historically grown accustomed to.

Throughout this chapter, we've reviewed the Plan Production capabilities and tools available to us within Civil 3D, where we can automatically generate sheet files of our design, leading to a seamless approach to submitting our design as part of the Digital Delivery process. As we've witnessed in the past, the sheeting process has begun much earlier on in the design process, typically just before a Conceptual Design review.

However, as the modeling process becomes more intricate, and technology advancements continue to increase the possibilities of reviewing our models in place of sheets, we must spend more time and effort upfront ensuring that our models are truly representative of our design and that we are maintaining dynamic linking between as many modeled objects as possible to minimize risk and rework later on down the road.

This chapter marks the end of this introductory guide for Civil 3D for infrastructure analysis and design. We have traversed every aspect of Civil 3D from grading and corridor modeling to pipe networks and production plan layouts. Civil 3D is a powerful tool that can make your civil designs not only easier to generate but also more intelligent and dynamic as projects evolve.

This guide is a practical walk-through of the real-world tools you will use every day for active team projects. As you wrap up these exercises, you should have foundational skills to leverage far into your design career. Civil 3D has many more tools, integrations, and tricks for carrying your work into construction, coordination, and even further after the design has been finalized. Use this guide as a tool to plant your feet in the civil software design world and look for more resources to take you to the next level.

Index

Other Books You May Enjoy

If you enjoyed this book, you may be interested in these other books by Packt:

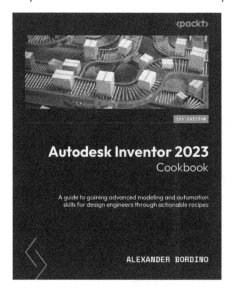

Autodesk Inventor 2023 Cookbook

Alexander Bordino

ISBN: 978-1-80181-050-0

- Build upon the fundamentals of parts, assemblies, and drawings
- Understand how to use advanced modeling tools such as iFeatures, iLogic, and more
- Develop your experience with parametric design methodologies
- Explore surface modeling and project management techniques
- Design efficiently with design accelerators to drive automation
- Understand and apply Finite Element Analysis

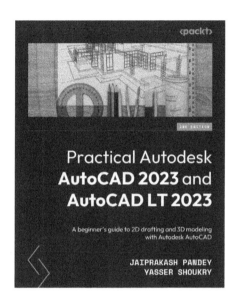

Practical Autodesk AutoCAD 2023 and AutoCAD LT 2023 - Second Edition

Jaiprakash Pandey , Yasser Shoukry

ISBN: 978-1-80181-646-5

- Understand CAD fundamentals like functions, navigation, and components
- Create complex 3D objects using primitive shapes and editing tools
- Work with reusable objects like blocks and collaborate using xRef
- Explore advanced features like external references and dynamic blocks
- Discover surface and mesh modeling tools such as Fillet, Trim, and Extend
- Use the paper space layout to create plots for 2D and 3D models
- Convert your 2D drawings into 3D models

Packt is searching for authors like you

If you're interested in becoming an author for Packt, please visit authors.packtpub.com and apply today. We have worked with thousands of developers and tech professionals, just like you, to help them share their insight with the global tech community. You can make a general application, apply for a specific hot topic that we are recruiting an author for, or submit your own idea.

Share Your Thoughts

Now you've finished *Autodesk Civil 3D from Start to Finish*, we'd love to hear your thoughts! Scan the QR code below to go straight to the Amazon review page for this book and share your feedback or leave a review on the site that you purchased it from.

https://www.amazon.in/review/create-review/error?asin=1803239069

Your review is important to us and the tech community and will help us make sure we're delivering excellent quality content.

Download a free PDF copy of this book

Thanks for purchasing this book!

Do you like to read on the go but are unable to carry your print books everywhere?

Is your eBook purchase not compatible with the device of your choice?

Don't worry, now with every Packt book you get a DRM-free PDF version of that book at no cost.

Read anywhere, any place, on any device. Search, copy, and paste code from your favorite technical books directly into your application.

The perks don't stop there, you can get exclusive access to discounts, newsletters, and great free content in your inbox daily

Follow these simple steps to get the benefits:

1. Scan the QR code or visit the link below

https://packt.link/free-ebook/9781803239064

2. Submit your proof of purchase

3. That's it! We'll send your free PDF and other benefits to your email directly

www.ingramcontent.com/pod-product-compliance
Lightning Source LLC
Chambersburg PA
CBHW081502050326
40690CB00015B/2895